International Conference on Combustion Physics and Chemistry (ComPhysChem'18)

MATEC Web of Conferences Volume 209 (2018)

Samara, Russia
24 – 28 July 2018

Editors:

A. Mebel
V. Azyazov

ISBN: 978-1-5108-7251-6

Printed from e-media with permission by:

Curran Associates, Inc.
57 Morehouse Lane
Red Hook, NY 12571

Some format issues inherent in the e-media version may also appear in this print version.

Printed by Curran Associates, Inc. (2018)

For additional information, please contact EDP Sciences – Web of Conferences at the address below.

EDP Sciences – Web of Conferences
17, Avenue du Hoggar
Parc d'Activité de Courtabœuf
BP 112
F-91944 Les Ulis Cedex A
France

Phone: +33 (0) 1 69 18 75 75
Fax: +33 (0) 1 69 28 84 91

contact-edps@webofconferences.org

Additional copies of this publication are available from:

Curran Associates, Inc.
57 Morehouse Lane
Red Hook, NY 12571 USA
Phone: 845-758-0400
Fax: 845-758-2633
Email: curran@proceedings.com
Web: www.proceedings.com

TABLE OF CONTENTS

ComPhysChem'18

International Conference
Combustion Physics and Chemistry - ComPhysChem'18
Samara, Russia, 24-28 July

This collection comprises selected papers presented at the International Conference on Combustion Physics and Chemistry (ComPhysChem'18) held at Samara National Research University on July 24-27, 2018 (Samara, Russia). The papers cover a broad range of fundamental and applied topics in the area of combustion physics and chemistry including fundamental physical-chemical processes, quantum chemical studies of potential energy surfaces of chemical reactions in flames, kinetics and dynamics of elementary processes, mathematical modeling of processes in combustion, laser and optical diagnostics, chemical, plasma, and laser initiation of combustion, flame structure, formation and destruction of polycyclic aromatic hydrocarbons (PAH), soot, graphene, and carbonaceous nanoparticles, chromatography and measurement methods, and ecological issues related to combustion.

The Conference was supported by the Russian Foundation for Fundamental Research (Grant No. 18-03-20026) and the Ministry of Education and Science of the Russian Federation (Grant No. 14.Y26.31.0020).

Alexander Mebel
Valeriy Azyazov

Organized by:

SAMARA UNIVERSITY

P. N. Lebedev Physical Institute of the Russian Academy of Sciences edp sciences

Web of Conferences

Statement of Peer review

In submitting conference proceedings to *Web of Conferences*, the editors of the proceedings certify to the Publisher that

1. They adhere to its **Policy on Publishing Integrity** in order to safeguard good scientific practice in publishing.
2. All articles have been subjected to peer review administered by the proceedings editors.
3. Reviews have been conducted by expert referees, who have been requested to provide unbiased and constructive comments aimed, whenever possible, at improving the work.
4. Proceedings editors have taken all reasonable steps to ensure the quality of the materials they publish and their decision to accept or reject a paper for publication has been based only on the merits of the work and the relevance to the journal.

Title, date and place of the conference The International Conference on Combustion Physics and Chemistry, 24-28 July, 2018, Samara, Russia

Proceedings editor(s): Alexander M. Mebel
Valeriy N. Azyazov

Date and editor's signature

Methane-air flame thermometry using planar lase-induced fluorescence (PLIF)

Victor Fabelinsky[1], *Vitaly* Kobtsev[2*], *Sergey* Kostritsa[2], *Dimitrii* Kozlov[1,2], *Valerii* Smirnov[1,2], and *Konstantin* Vereschagin[1,2]

[1]A.M. Prokhorov General Physics Institute, Russian Academy of Sciences, 119991 Moscow, Russia
[2]Central Institute of Aviation Motors, 111116 Moscow, Russia

Abstract. Temperature measurements were performed in a central cross-section of a partially premixed atmospheric methane-air flame of a laboratory burner using two-line planar laser-induced fluorescence (PLIF) of OH radicals and coherent anti-Stokes Raman scattering (CARS) by H_2 molecules. The results of a comparative analysis of 2D PLIF and single-point CARS thermometry of the flame are presented.

1 Introduction

Methods of linear and nonlinear laser spectroscopy are widely used in combustion research for aerospace engineering [1] to measure gas temperature and species concentration. In particular, the planar laser-induced fluorescence (PLIF) technique was applied [2-4] to hydrogen-containing flame diagnostics for detection of the hydroxyl (OH) radical concentration and temperature spatial distributions. These data allow a more detailed flame characterization and can be used for verification of the combustion models.

Strong UV rovibronic transitions of the OH molecule $A^2\Sigma^+ \leftarrow X\,^2\Pi$ bands provide a possibility to excite and detect 2D PLIF signals in the range with no IR- and visible flame radiation. On the other hand, CARS spectroscopy enables precise high spatial resolution single-point measurements. Therefore, PLIF and CARS techniques complement each other in flame diagnostics.

The present work is devoted to the experimental realization of spatially-overlapped registration of 2D PLIF images of OH radicals, excited at two wavelengths, and a set of single-point CARS-spectra of H_2 molecules in a partially-premixed atmospheric methane-air flame of a laboratory burner, and to a comparative analysis of flame temperature distributions provided by these methods. The analysis is aimed to clarify the applicability of the two-line PLIF technique to thermometry of a spatially-inhomogeneous gas flow in a flame.

* Corresponding author: kobtsev.vitaly@gmail.com

2 Experimental

The measurements were performed using a laboratory setup that includes the PLIF and CARS diagnostic systems, and a slot Wolfhard-Parker burner, with a "flat" flow of a combustible mixture in the center and a co-flow of air from aside, similar to that described in [5]. PLIF of OH intensity distributions, at two excitation wavelengths, and single-point CARS-spectra of H_2 molecules were being recorded in the same central cross-section of a stationary partially-premixed atmospheric methane-air flame of the burner. This cross-section is perpendicular to the flame symmetry plane, i.e. to the slot of the burner.

The PLIF system consists of a tunable narrowband ($\approx$ 0.05 cm^{-1}) 10-ns 8-Hz frequency-doubled dye laser pumped by the second-harmonic radiation of a Nd:YAG laser, an optical system which forms a flat "laser knife", and an intensified CCD camera equipped with a UV objective and an interference (central wavelength 310 nm, HWHM 10 nm) and a colored glass short-pass (cut-off wavelength 350 nm) filters which block the excitation and visible radiation. To obtain temperature distributions, PLIF signals were excited using two rovibronic transitions of the $A\,^2\Sigma^+ \leftarrow X\,^2\Pi\,(1,0)$ band of OH radicals ($Q_1(3)$ – 35420.38 cm^{-1}, and $Q_2(10)$ – 35121.08 cm^{-1}), with the energies of the lower states $E_1 = 202.38$ cm^{-1} and $E_2 = 2056.55$ cm^{-1}, respectively. Images of PLIF in the $X\,^2\Pi \leftarrow A\,^2\Sigma^+$ (0, 0) band for each of the excitation wavelengths, $I_1(m,n)$ and $I_2(m,n)$, where m and n are the image pixel coordinates, were obtained at about 1 mJ of the incident laser pulse energy as accumulation of 200 consecutive CCD frames.

If the excitation laser intensity distributions at two lines can be assumed to be homogeneous within the imaged spatial area, the ratio of the intensities in the PLIF images

$$R(m,n) = I_2(m,n)/I_1(m,n) \qquad (1)$$

can be expressed as[3]

$$R(m,n) = C\times\exp[-(E_2 - E_1)/(kT(m,n))] \qquad (2)$$

Here, k is the Boltzmann constant, $T(m,n)$ is local gas temperature, and C is a coefficient. This coefficient depends on the ratios: of the absorption cross-sections of the involved transitions, of the excitation pulse energies and of the overlaps of the excitation and absorption line profiles, as well as of the parts of the excited $A\,^2\Sigma^+(v = 1)$ OH molecules transferred into the upper fluorescing $A\,^2\Sigma^+(v = 0)$ state. If the assumption is valid that the coefficient C weakly depends on gas temperature and composition, it can be derived experimentally by comparing the value of R in a particular spatial point to the temperature in this point, independently determined using a reference technique. Further on, temperatures in all other spatial points can be calculated.

In this work, vibrational CARS thermometry was used as the reference technique to derive and compare the values of C in a few spatial points, checking the validity of the assumption of its weak variation in the PLIF-probed cross-section of the flame, and to obtain 2D temperature distributions based on the PLIF signals in the areas where this assumption is valid. CARS-spectra of H_2 molecules, which appear in the fuel-rich zones of the flame, were recorded using the CARS system employing broadband Stokes radiation and non-collinear configuration of crossing the beams. The detailed description of the system is given in [6]. The probe volume was ~ 0.05 mm in diameter and ~ 2 mm in length. CARS-spectra were recorded across the regions of the highest temperature gradients before the flame fronts at a few heights, Z, above the edge of the burner.

3 Results and discussion

The atmospheric partially-premixed methane-air flame (Fig. 1, at the left) was operating at 5.9 slpm of methane-air mixture ($\Phi = 2$) and 22.3 slpm of dry air. PLIF signal distributions (Fig. 1, in the center) were recorded in the central cross-section of the flame.

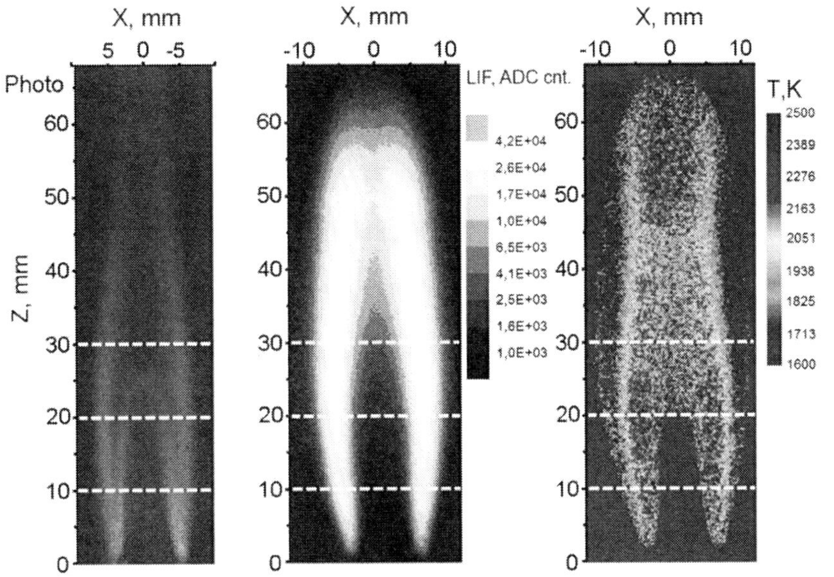

Fig. 1. A photograph of the partially-premixed fuel-rich atmospheric methane-air flame of the Wolfhard-Parker burner along the slot (at the left), a PLIF of OH intensity distribution excited via the $Q_2(10)$ transition (in the center), and the calculated temperature distribution in the central cross-section of the flame (at the right) on the same scale; *X* is the distance from the flame symmetry plane. The dashed lines show the heights above the edge of the burner at which CARS temperature measurements involved in the derivation of the coefficient *C* were performed.

CARS-spectra of H_2, recorded in the same cross-section at the heights 10, 20, and 30 mm above the edge of the burner with steps of 1-2 mm, were processed to derive local gas temperatures. The obtained values are in a reasonable agreement with those derived from vibrational CARS-spectra of N_2 in the similar points of the methane-air flame of a similar Wolfhard-Parker burner run at the equivalent flow conditions [7].

The values of the coefficient *C* of correspondence between the ratios of PLIF intensities and the temperatures were calculated for the points of CARS measurements. In the significant part of the probed flame cross-section area, where OH radicals exist and fluoresce, the spread of the derived *C* values was found to be less than 10%. This confirmed the validity of the assumption of the weak dependence of ***C*** on temperature and species concentration for the considered part of the cross-section and allowed us to use the mean value of *C* for the subsequent calculation of the 2D temperature distribution (Fig. 1, in the right). For this, the PLIF signal cut-off level was chosen to be 5% of the maximal value. This selection was conditioned by the amplitude of noise of the PLIF signals resulting in significant errors in derived temperatures if the values below this level were taken.

Figs. 2 and 3 represent spatially averaged over 2 mm in height profiles of the calculated 2D temperature distribution, together with CARS temperature data, obtained in the present work and in [7], at the heights of 10 and 30 mm above the burner edge. The profiles of the PLIF signal distribution in Fig. 1, at the same heights, are also presented. The spatial filtering of the PLIF signals is required to obtain smooth temperature distributions at the relatively high spread of the derived values. As a result, the temperature measurements using PLIF demonstrate lower spatial resolution than those employing CARS, but provide larger field of data for a shorter time.

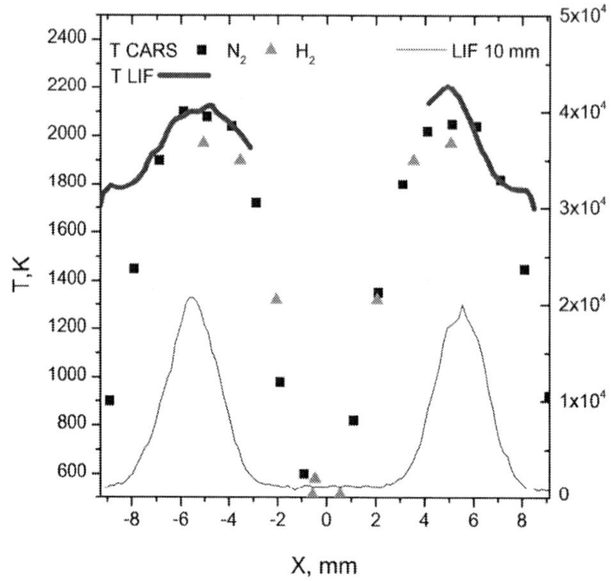

Fig. 2. Calculated temperature profiles (smoothed OH PLIF measurements – bold red line, CARS by H_2 – green triangles, CARS by N_2 [7] – black squares) and PLIF intensity profile (thin black line) at 10 mm above the burner edge.

The defined spread of C values does not result in significant deviations between the PLIF and CARS temperature profiles (less than 150 K), and the values provided by the techniques are in a reasonable agreement. Note, that the PLIF signals are not intense enough for the calculations in the low-temperature regions ($T < 1600$ K) - in the central part of the combustible mixture flow, at the heights below 20 mm (the inner flame front, Fig, 1), and in the outer part of the flame.

3 Conclusion

Complementary measurements were carried out in the central section of the partially-premixed atmospheric methane-air flame of the laboratory slot burner with spatially-inhomogeneous species and temperature profiles using two-line PLIF of OH radicals and CARS by H_2 molecules.

Processing of the PLIF images registered at two excitation wavelengths employing the results of CARS temperature measurements allowed us to derive the values of the correspondence coefficient C between the ratios of PLIF intensities and the temperatures in the specific points of the PLIF images.

Small spread of the derived C values was obtained in the significant part of the probed flame cross-section area, where OH radicals exist and fluoresce. This confirmed the validity of the assumption of the weak dependence of C on temperature and species concentration for the considered part of the cross-section and allowed us to convert the two PLIF images into the 2D temperature field. The calculated temperatures were found to match the results of our H_2-CARS measurements and N_2-CARS data from the literature.

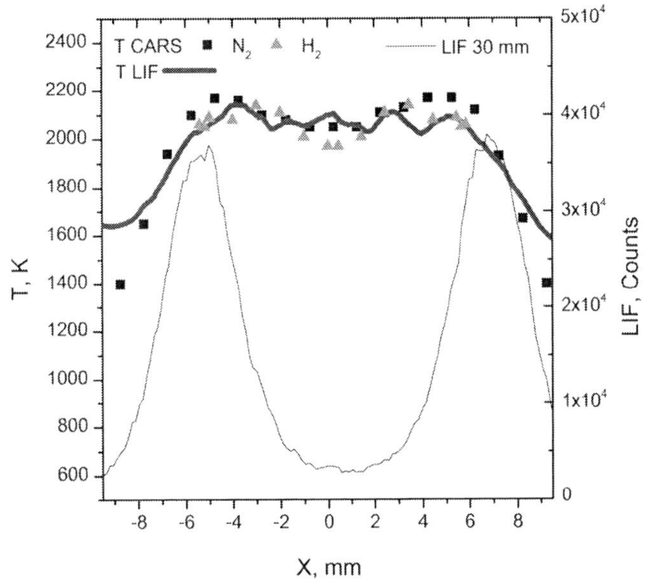

Fig. 3. Calculated temperature profiles (smoothed OH PLIF measurements – bold red line, CARS by H_2 – green triangles, CARS by N_2 [7] – black squares) and PLIF intensity profile (thin black line) at 30 mm above the burner edge.

The help of T. Seeger (University of Siegen, Germany) with manufacturing and tests of the Wolfhard-Parker burner in frame of the Cooperation Program is gratefully acknowledged. The work was financially supported by CIAM Foundation for Priority Research.

References

1. A. Ehn, J. Zhu, X. Li, and J. Kiefer, Appl. Spectrosc. **71** (5), 1-26, (2017)
2. R.K. Hanson, K. Ronald, J. M. Seitzman, P. H. Paul. Appl. Phys. B, **50** (6), 441-454 (1990)
3. J. M. Seitzman, R. K. Hanson, P. A. DeBarber, C. F. Hess, Applied Optics, **33** (18), 4000-4012 (1994)
4. U. E. Meier, D. Wolff-Gaßmann, W. Stricker, Aerosp. Sci. and Tech. **4** (6), 403 - 414 (2000)
5. K.C. Smyth, J.H. Miller, R.C. Dorfman, W.G. Mallard, R.J. Santoro Combust. Flame, **62**, 157-181 (1985)
6. V.V. Smirnov, S.A. Kostritsa, V.D. Kobtsev, N.S. Titova, A.M. Starik, Combust. Flame, **162** (10), 3554–3561 (2015)
7. A. Datta, F. Beyrau, T. Seeger, A. Leipertz, Combust. Sci. and Techn. **176**, 1-20 (2004)

Three-dimensional simulation of combustion, detonation and deflagration to detonation transition processes

Nickolay Smirnov[1,2,*], *Valeriy* Nikitin[1,2]

[1] Moscow M.V. Lomonosov State University, Moscow 119991, Leninskie Gory, 1, Russian Federation
[2] Federal Science Center "Scientific Research Institute for System Analysis of Russian Academy of Sciences", Moscow 117218, Nakhimovskiy, 36-1, Russian Federation

Abstract. The paper presents results of numerical and experimental investigation of mixture ignition and detonation onset in shock wave reflected from inside a wedge. Contrary to existing opinion of shock wave focusing being the mechanism for detonation onset in reflection from a wedge or cone, it was demonstrated that along with the main scenario there exists a transient one, under which focusing causes ignition and successive flame acceleration bringing to detonation onset far behind the reflected shock wave. Several different flow scenarios manifest in reflection of shock waves all being dependent on incident shock wave intensity: reflecting of shock wave with lagging behind combustion zone, formation of detonation wave in reflection and focusing, and intermediate transient regimes. Comparison of numerical and experimental results made it possible to validate the developed 3-D transient mathematical model of chemically reacting gas mixture flows incorporating hydrogen – air mixtures.

1 Introduction

Self-sustaining waves can propagate in meta-stable media; energy needed to support such waves is released by the wave itself. As a rule, two regimes of propagation exist, subsonic and supersonic; the difference is based on the different mechanisms of medium activation. Processes of transition between those regimes are less studied up to now, in comparison with pure subsonic or supersonic modes. Control of detonation onset is necessary in perspective pulse detonation engines using hydrogen-air mixtures in the working cycle, which are under development now. In our studies we'll use hydrogen fuel because, on one hand, it is a very perspective fuel making the engine exhaust much cleaner than that for hydrocarbon combustion [1,2], and on the other hand, chemical kinetics for hydrogen – air mixtures combustion are well developed [3-8]. The advantages of a constant volume combustion cycle as compared to constant pressure combustion in terms of thermodynamic

* Corresponding author: ebifsun1@mech.math.msu.su

efficiency has focused the search for advanced propulsion on detonation engines [9-10]. The thermodynamic efficiency of Chapman – Jouguet detonation as compared with slow combustion modes is due to the minimal entropy of the exhaust jet [10]. The present investigation is focused on initiation of detonation within limited in size spatial structures by relatively weak initiators using advantage of shock waves focusing and energy cumulation.

2 Mathematical model

The mathematical model for simulations of multi-component gas dynamics with chemical reactions including transport phenomena effects and turbulence was described in details in [11]. It contains the species mass, momentum and energy variation equations, which account for turbulent fluxes based on two-equation turbulence model of Wilcox k-omega [12].

$$\frac{\partial \rho_k}{\partial t} + \frac{\partial}{\partial x_j}\left(\rho_k u_j - J_{kj}\right) = \dot{\omega}_k \tag{1}$$

$$\frac{\partial \rho u_i}{\partial t} + \frac{\partial}{\partial x_j}\left(\rho u_i u_j\right) + \frac{\partial p}{\partial x_i} = \frac{\partial \tau_{ij}}{\partial x_j} \tag{2}$$

$$\frac{\partial E}{\partial t} + \frac{\partial}{\partial x_j}\left((E + p)u_j - J_{Q,j}\right) = \frac{\partial u_i \tau_{ij}}{\partial x_j} \tag{3}$$

In equations Eq.(1) – (3), index k takes values 1 ... N_C (number of components), and indices i,j – values 1, 2, 3 (number of dimensions); repeated indices presume summation. In total, there are N_C +4 differential equations in the set. There is summation upon the repeating indices in Eq. (1) – (3).

Differential equations (1) – (3) are complemented with algebraic relations and algebraic representations for chemical and mass and energy sources [12].

Explicit second-order in space and time method based on the MUSCL-interpolation of variables on a face at a convective flux calculation. Interpolation direction choice and pressure interpolation were performed by means of AUSMP method. The method was implemented on a regular grid containing equal elements (rectangular parallelepipeds) connected generally in arbitrary topology. The source code was written in C. Parallel execution support was implemented using OpenMP library.

3 Results of numerical simulations

The problem simulating processes in the final section of the cylindrical shock tube was regarded. The geometry and size of computational domain coincides with that in the shock tube used in experiments. The shock tube diameter was $2R$=76 mm, length – L=720 mm. (Fig. 1).

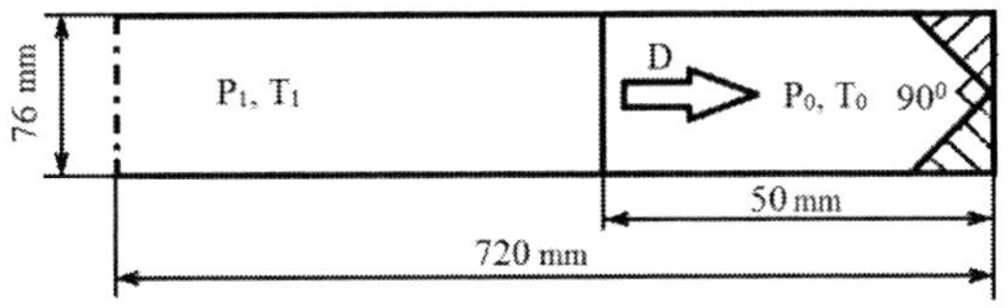

Fig. 1. Scheme of computational domain, initial and boundary conditions.

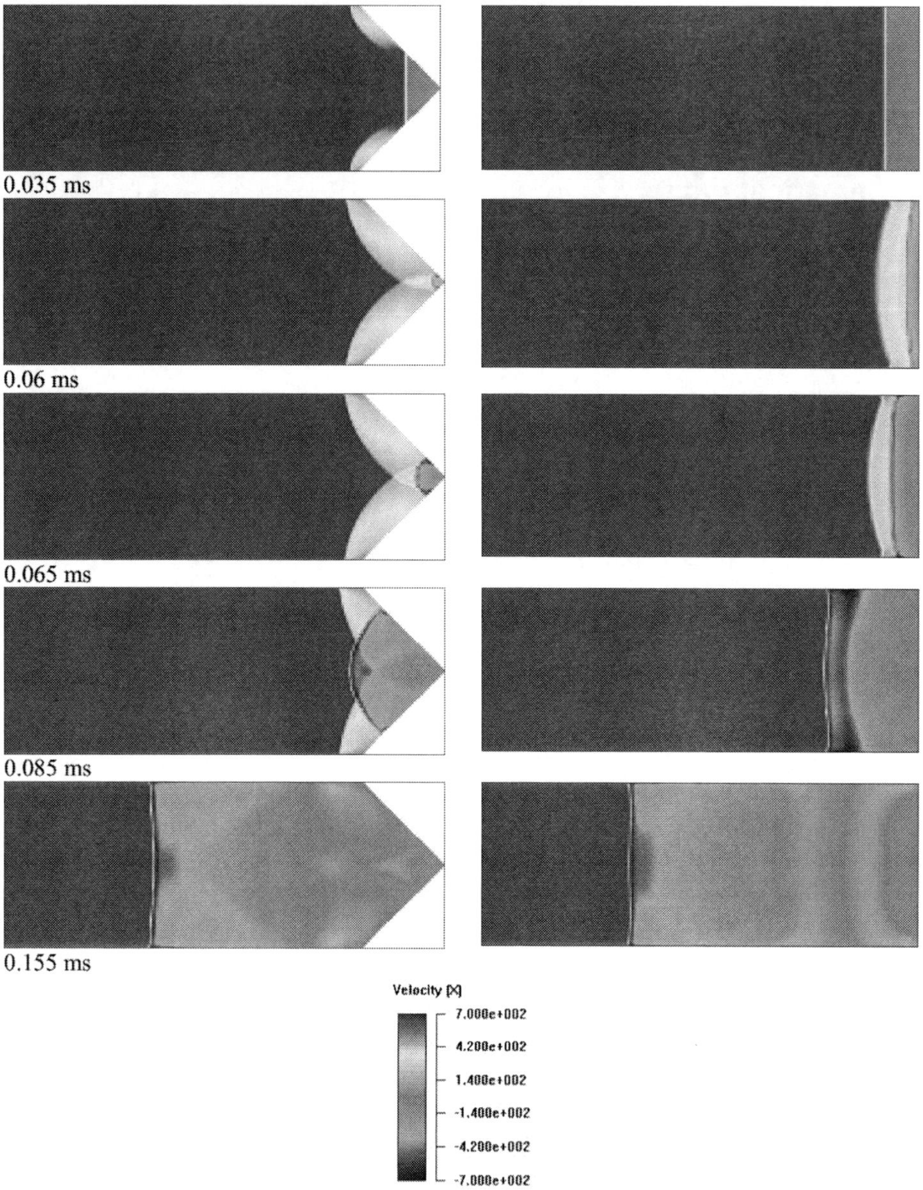

Fig. 2. Axial velocity fields in (m/s) on focusing of a strong shock wave and detonation onset in reflection from a wedge surface. Meridian cross-sections Oxy (left) and Oxz (right).

At time 0.065 ms onset of detonation is observed in the tip of the wedge. Then detonation wave propagates outside the wedge. Temperature maps in Oxy and Oxz planes testify that the burnt gas has temperature 1600 – 2000 K. For smaller intensity of incident shock wave its focusing in reflection from a wedge does not bring to immediate formation of detonation wave but it could cause mixture ignition and flame propagation. A transient regime is possible, characterized by formation of a leading reflected shock wave and flame zone lagging behind it. Then acceleration of the flame could bring to onset of detonation.

Fig. 3 illustrates successive stages of the process of deflagration to detonation transition after reflecting of shock wave from a wedge

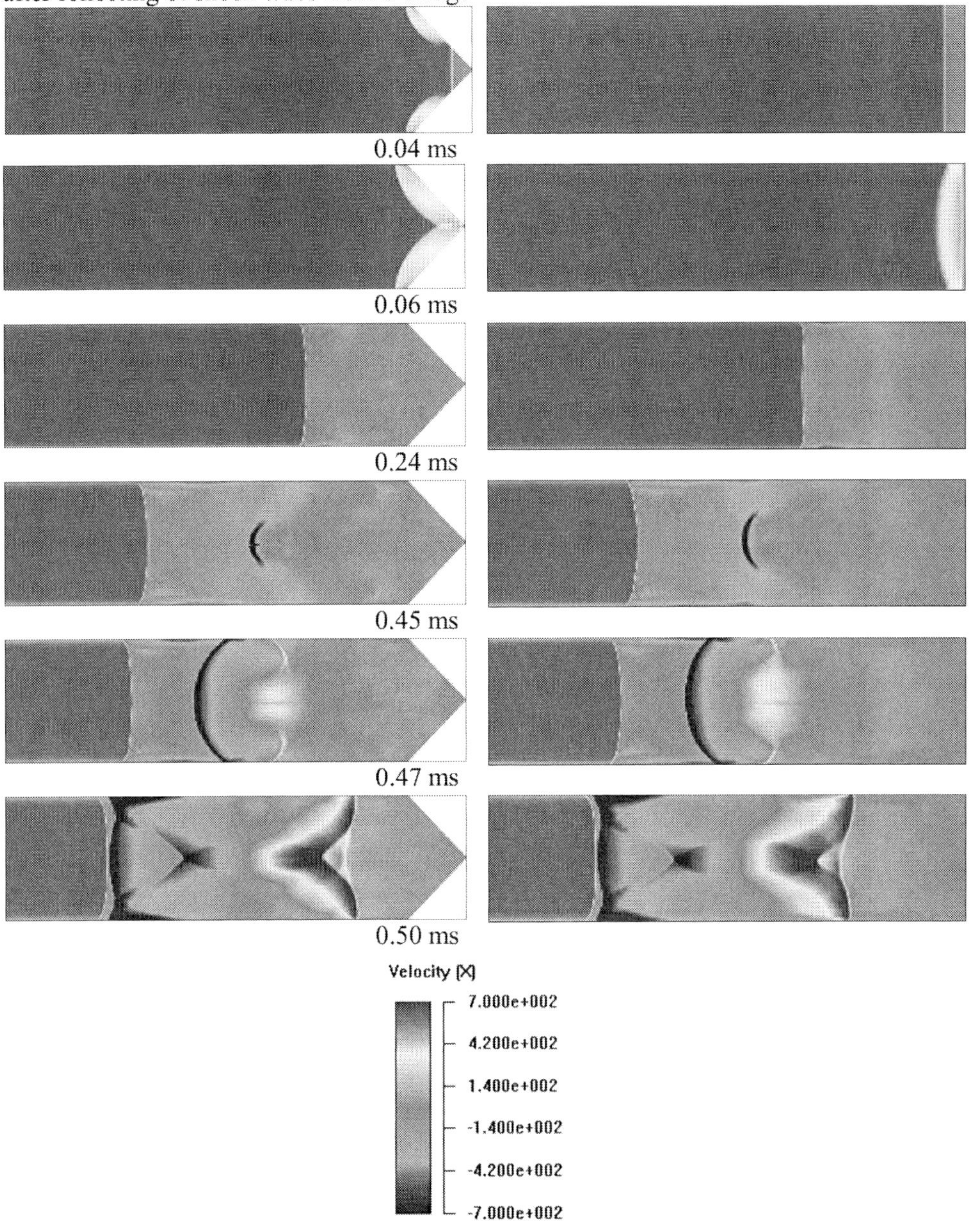

Fig. 3. Axial velocity fields in (m/s) on focusing of a weak shock wave and deflagration to detonation transition in reflection from a wedge surface. Meridian cross-sections Oxy (left) and Oxz (right).

Velocity field in the Fig. 3 testifies that on reflection of shock wave from the wedge surface cumulative effect is not strong enough for the onset of detonation wave. Shock wave is reflected. Ignition of mixture in the tip takes place later, and accelerated flame moving in pre-compressed and non-uniformly heated gas in the long run gives birth to local explosion and formation of detonation and retonation waves. Velocity field makes it possible distinguishing between detonation and retonation waves due to positive and negative velocities being depicted in different colours.

4 Conclusions

It was demonstrated that onset of detonation due to focusing of a strong shock wave takes place via an overdriven detonation mode. Then detonation wave slows down to a self-sustaining mode.

The transient regime has the following characteristic stages. First, at the tip of the wedge in the center a hot spot appears, which later increases rapidly in all directions, nevertheless lagging behind the reflected shock wave. Combustion wave is thus formed. Second, the combustion zone grows the leading front being unstable. Wrinkles are formed on the leading front, which is noticeable especially in the plane orthogonal to wedge. Third, velocity of combustion wave drastically increases, especially in the wrinkled zone, the wrinkles become deeper. Forth, the zone grows near the axis, which testifies formation "explosion in the explosion" and onset of detonation wave. The detonation wave begins reflecting from side walls of the cylinder and further propagate as detonation and retonation waves in all directions.

Present study was supported by the program of Russian Academy of Science « Algorythmic sttling and codes for simulation of multiscale processes and combustion" (№ 0065-2018-0007)

References

[1] CHEMKIN. A software package for the analysis of gas-phase chemical and plasma kinetics. CHE-036-1. Chemkin collection release 3.6. Reaction Design, September 2000.
[2] N.M. Marinov, W.J. Pitz, C.K. Westbrook, M. Hori, N. Matsunaga, An Experimental and Kinetic Calculation of the Promotion Effect of Hydrocarbons on the NO-NO_2 Conversion in a Flow Reactor. Proceedings of the Combustion Institute, Volume 27, pp. 389-396, 1998. (UCRL-JC-129372). UCRL-WEB-204236.
[3] R.J. Kee, J.A. Miller, and T.H. Jefferson. Chemkin: a general-purpose, problem-independent, transportable Fortran chemical kinetics code package. Sandia National Laboratories Report SAND80-8003 (1980).
[4]. S. Browne, J. Ziegler, and J. E. Shepherd. Numerical Solution Methods for Shock and Detonation Jump Conditions. GALCIT Report FM2006.006 July 2004-Revised August 29, 2008.
[5] S. Gordon, and B.J. McBride. Computer Program for Calculation of Complex Chemical Equilibrium Compositions and Applications I. Analysis. NASA RP-1311, October 1994.
[8] U. Maas, J. Warnatz. Ignition process in hydrogen-oxygen mixtures. Combustion and Flame, 74, No. 1, 1988, pp. 53-69.
[9] N.N. Smirnov, V.F. Nikitin. Modeling and simulation of hydrogen combustion in engines. International Journal of Hydrogen Energy. (2014), vol. 39, Iss.2, 1122-1136. http://dx.doi.org/10.1016/j.ijhydene.2013.10.097
[10] N.N. Smirnov, V.B. Betelin, V.F. Nikitin, Yu.G. Phylippov, Jaye Koo. Detonation engine fed by acetylene–oxygen mixture, Acta Astronautica (2014), vol. 104, 134-146. http://dx.doi.org/10.1016/j.actaastro.2014.07.019i
[11] Smirnov N.N., Penyazkov O.G., Sevrouk K.L., Nikitin V.F., Stamov L.I., Tyurenkova V.V. Detonation onset following shock wave focusing. Acta Astronautica, 2017. Volume 135, Pages 114-130.
[12] D.C. Wilcox. Turbulence modeling for CFD. DCW Industries, Inc. 1994.

MATEC Web of Conferences **209**, 00004 (2018) https://doi.org/10.1051/matecconf/201820900004
ComPhysChem'18

Numerical Implementation and validation of turbulent premixed combustion model for lean mixtures

Siva Muppala [1,1,] and *Vendra C.* Madhav Rao [2]

[1] Faculty of Science, Engineering & Computing, Kingston University London, SW15 3DW, UK
[2] Department of Engineering, University of Warwick, Coventry, CV4 7AL

Abstract. The present paper discusses the numerical investigation of turbulent premixed flames under lean conditions. Lean premixed combustion, a low NOx emission technique but are prone to instabilities, extinction and blow out. Such flames are influenced by preferential diffusion due to different mass diffusivities of reactants and difference between heat and mass diffusivities in the reaction zone. In this numerical study, we estimate non-reacting flow characteristics with implementation of an Algebraic Flame Surface Wrinkling Model (AFSW) in the open source CFD code OpenFOAM. In these flows, the mean velocity fields and recirculation zones were captured reasonably well by the RANS standard k-epsilon turbulence model. The simulated turbulent velocity is in good agreement with experiments in the shear-generated turbulence layer. The reacting flow study was done at three equivalence ratios of 0.43, 0.5 and 0.56 to gauge the ability of numerical model to predict combustion quantities. At equivalence ratios 0.5 and 0.56 the simulations showed numerical oscillations and non-convergence of the turbulent quantities. This leads to a detailed parametric variation study where, the pre-constant of AFSW model is varied with values 0.3, 0.35 and 0.4. However the study revealed the weak dependence of pre-constant value on the equivalence ratio. Hence the pre-constant value is fit for specific equivalence ratio based on the parametric variation study. The tuned AFSW model with fitted pre-constant specific to given equivalence ratio predicted are compared with experiments and discussed. The tuned AFSW model produced turbulent flame speed values which are good agreement with experiments.

1. Introduction

The study of premixed turbulent combustion is important to make the combustion more efficient and less harmful for nature. The research work focuses on the behaviour of lean turbulent premixed flame at high pressures. A well compiled large set of experimental data is available for validation. This data is chosen since the two main quantities of turbulent premixed combustion, turbulent flame speed (S_T) and flame brush thickness (δ_T) have been measured successfully. The specific objective of the current study is to make contribution in the field of numerical modelling of turbulent premixed combustion by implementing and validating a universal combustion model. The numerical simulations are carried out using RANS approach for modelling turbulence and laminar flamelet approach for modelling premixed combustion. The laminar flamelet modelling assumes that combustion happens in thin reaction zone regimes wherein the flame locally behaves as a laminar flame and turbulent flow structures are much bigger than combustion mechanism structures; hence the effect of turbulence on combustion is to wrinkle the local laminar flame and thereby enhance the combustion through increased flame surface area. The numerical simulations are carried out using open source CFD tool OpenFOAM solver.

A large set of well compiled experimental results have been produced through experiments carried out at Paul Scherrer Institute (Baden), Switzerland [1, 2, and 3]. The premixed turbulent flame is studied in a high-pressure combustion test rig (shown in figure 1, below) that is designed for a maximum operating pressure of 30 bars and a maximum air flow rate of 750m^3/hr. The combustion air/fuel mixture can be preheated up to 823K with an electrical heater. The study highlighted characteristics of non-reacting turbulent flow field and typical high-pressure turbulent premixed flame. The PIV measurements of flow field and planar Laser Induced fluorescence of OH radical distribution were used. The operating conditions for non-reacting flows (293K, 1bar) and for reacting flows a preheating temperature of 673K, pressures up to 14.44 bar (absolute), an equivalence ratio in the range ϕ=0.43–0.56 of premixed methane/air mixture at bulk flow velocity of 40 m/s. The turbulent intensity and integral length scale in the inlet are controlled using turbulence grid which is defined by hole diameter, blockage ratio and the axial position of turbulence grid within the combustor inlet section (see figure1). (for more details refer to Griebel et al. (2006)

[1] Corresponding author: S.Muppala@kingston.ac.uk

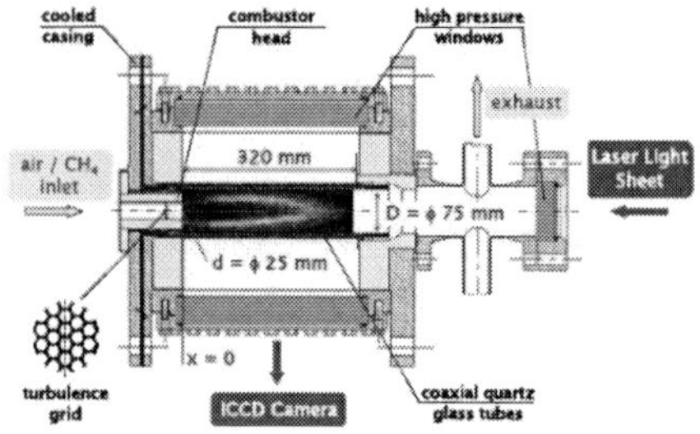

Fig 1. The experimental setup by Griebel et al. (2006)

2. Numerical Setup

The computational domain of combustion chamber used for all numerical simulations in this study consists of two co-axial cylinders of diameters 25mm which is a pre-inlet pipe and combustion chamber of diameter 75mm. The pre-inlet pipe of 25mm diameter goes for a length of 30mm serving two purposes. First, helps in letting the homogeneous turbulence flow enter the combustion chamber and second, creates a sudden expansion leading to recirculation zone which stabilises the flame. The combustion chamber of 75mm diameter goes for length of 320mm. The figure 2 below shows the schematic diagram of the computational domain considered.

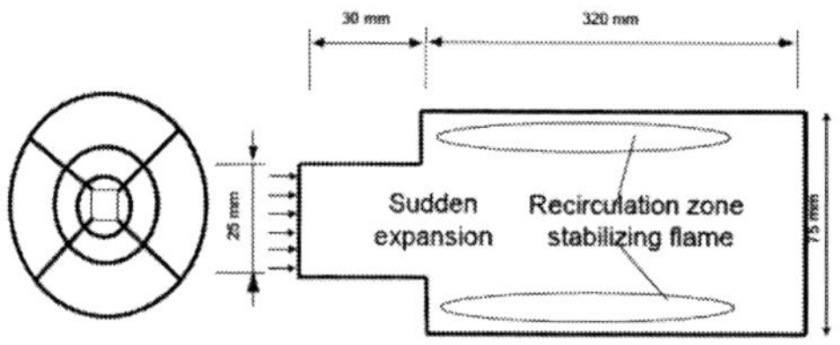

Fig. 2. The problem domain of High Pressure combustion chamber, (all dimensions in mm)

The computational grid for the above domain is created using ICEM [Ansys, Inc]. The grid independency test is carried out using two grids consisting of 300,000 cells and 600,000 cells. The grid with 600,000 cells is found to be adequate and all further reacting and non-reacting flow study is carried out using 600,000 cells grid.

3. Combustion Model

SPR Muppala (2005) model is a premixed turbulent combustion model which can be applied for different fuels up to pressure of 1MPa. Here the reaction source term of the reaction progress variable is modelled using laminar flame speed, corrected for stretch effects and turbulent flame surface area. The calculation of turbulent flame surface area is through a parameter called the flame-wrinkling ratio A_T/A . Flame wrinkling ratio is modelled with an algebraic parameterized relation. This term flame wrinkling ratio A_T/A is equivalent to turbulence flame wrinkling Ξ in Weller's Flame Area Combustion model [4]. The AFSW model is incorporated as the algebraic relation to calculate turbulence flame wrinkling as shown below. This model has been validated against various experimental data successfully [6, 7]. The combustion source term is modelled as,

$$\Sigma = \frac{A_T}{A} = \frac{S_T}{S_{L0}} = 1 + a\,\mathrm{Re}_t^{0.25}\left(\frac{u'}{S_{L0}}\right)^{0.3}\left(\frac{p}{p_0}\right)^{0.2}$$

Here, 'a' is called pre-constant with 0.46, this parameter is linked to Lewis number making it possible to study molecular effects of diffusion on combustion [7].

4. Results and discussion

Non-reacting flow: In the investigation of turbulent flow field characteristics of non-reacting flows, the approach has been to study grid independency and to choose the appropriate turbulence model. From the experiments it was not clearly understood the degree of turbulence level that can be used in numerical simulations. Hence a detailed numerical non-reacting flow study is carried to fix the proper turbulence level and turbulence model. The results have been presented in detail [8]. The discrepancy in the turbulent flow field prediction using two different grids, one with 300000 control volume cells and another with 600000 control volume cells is shown in figure 3. Clearly gird with 600000 cells showed a better match with the experiments.

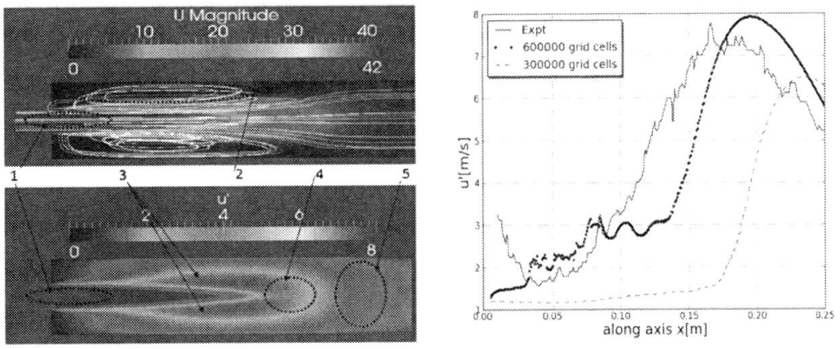

Fig. 3. Turbulent velocity u' along the axis of the combustion chamber for different grid cells with 5% turbulence level with the standard $k-\varepsilon$ turbulence model.

<u>Reacting flow study:</u> The RANS simulations were aimed at the study of behaviour of the turbulent premixed flames at different equivalence ratios, ϕ=0.43, 0.5 and 0.56. The flame characteristics like turbulent flame speed, flame position and flame brush thickness are measured and compared with experiments. The measurement of the flame length is the length along the axis of combustion chamber from start of the combustion chamber to the point where the value combustion regress variable is 0.5. The flame brush thickness is the distance along the axis of combustion chamber between the value combustion regress variable 0.05 to 0.95.

For ϕ=0.43: In the simulation study of turbulent premixed flames at varied equivalence ratios, the leanest mixture simulated was equivalence ratio φ=0.43. When such a lean mixture is under study low burning rates occur due to lower fuel concentrations. Under such lean conditions the flame is very sensitive to local flow changes or variations. The flame temperatures attained are low and combustion reaction happens to be an elongated long flame. The fig 3 gives contour plot of the reaction regress variable. The contour shows that a smooth elongated flame has been obtained, where the reactants indicated by colour red is getting burnt along flame (where the colour is changing from red to blue). The rest of the domain is covered by products (which is coloured in blue).

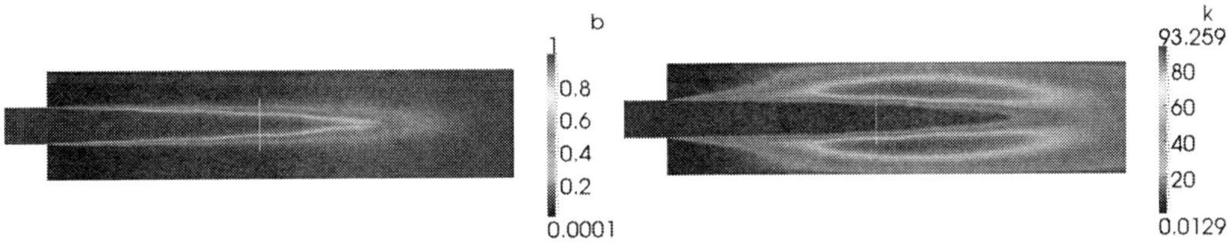

Fig 4. 'b' regress variable contour and 'k' turbulent kinetic energy contour for equivalence ratio 0.43 using AFSW combustion model with original pre-constant. *Colour convention* blue (b=0) is burnt or combustion gas and red (b=1) is un-burnt or inlet premixed gas with 0<b<1.

The quantitative comparisons of the flame parameters predicted by the simulation and experimental values have been shown in the table. The current model under-predicts flame position in relative to the experimental flame position. While the turbulent flame brush thickness is over predicted, the flame temperature values are in good agreement with theoretical adiabatic flame temperature values.

For ϕ=0.5: Due to lower equivalence ratio resulting in lower chemical burning rate in the previous case of simulation discussed, the chemical reaction zone predominantly prevails far downstream of the domain. The increment in the equivalence ratio from 0.43 to 0.5 leads to higher chemical burning rate leading to the propagation of chemical reaction zone towards the upstream of the domain. The density drop across the flame due to combustion causes an expansion wave in the flow. The flame obtained is more corrugated and short due to turbulent structures and chemistry influencing each other. The flame possesses a long flame brush thickness almost of the order of the flame length itself.

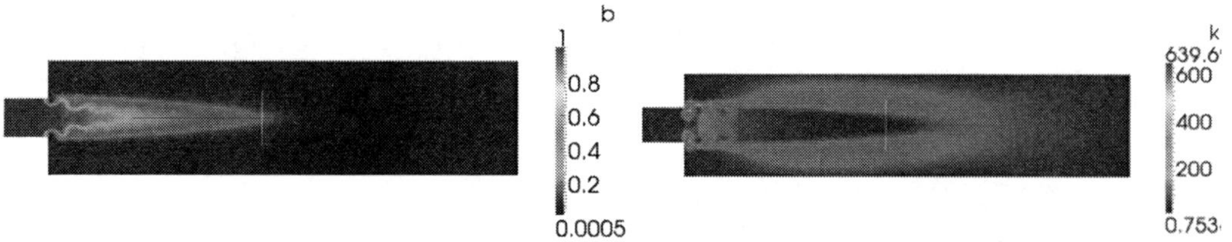

Fig 5. `b' regress variable contour and `k' turbulent kinetic energy contour for equivalence ratio 0.5 using AFSW combustion model with original pre-constant. Colour convention blue (b=0) is burnt or combustion gas and red (b=1) is un-burnt or inlet premixed gas with 0<b<1.

Turbulent flow field interaction with combustion is found to be very significant at equivalence ratio 0.5. Hence turbulence kinetic energy field did not converge even after 25 sweeps of domain. The turbulent kinetic energy contour shown below in figure 4 confirms the very non-uniform turbulent flow field.

For *ϕ=0.56*: The equivalence ratio has significant effect on the flame position and shape. The flame tends to shorten its length as the equivalence ratio is increased. The shape of the flame corresponds to the flame surface area. It is known from the fundamentals of chemical kinetics that, the rate of a reaction is directly proportional to the concentration of the reactants. Therefore at higher equivalence ratios the chemical burning rate increases due to higher fuel concentrations resulting in the higher temperatures and higher turbulent burning velocities.

Equivalence ratio, φ — Pre-constant = 0.3 — Pre-constant = 0.35 — Pre-constant = 0.4

φ=0.43

φ= 0.5

φ= 0.56

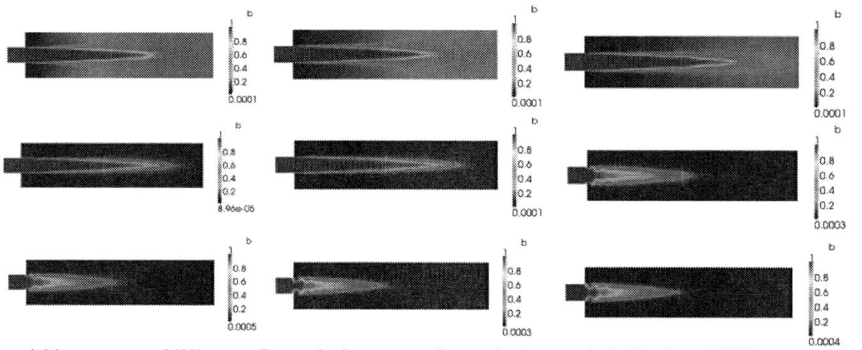

Fig 6. 'b' regress variable contour and 'U' mean flow velocity contour for equivalence ratio 0.56 using AFSW combustion model with original pre-constant.

Colour convention blue (*b*=0) is burnt and *b*=1 indicates unburned. As a parametric variation study to mimic experiments the pre-constant parameter from its original constant of 0.46 for methane/air flames in the AFSW model is varied from 0.3 to 0.4. The pre-constant parameter value is varied to study the significance of this parameter on the flame stability and characteristics. The values are brought in to observe the change in chemical reaction rate prediction and its effect on the turbulence and combustion interaction to see if a converged solution can be obtained.

The above set of figures in the table gives complete idea of flame behaviour at different pre-constant values for the AFSW model. It is observed that the pre-constant value predominantly affects the chemical burning rate prediction. It is also observed that pre-constant value is weakly dependent on the equivalence ratio, because at equivalence ratio 0.43 the mixture does not burn completely as the pre-constant value is decreased. At the same time, the decrease in the pre-constant value brought converged solution at equivalence ratio 0.5. This indicates that a universal value of pre-constant cannot be used for all equivalence ratios.

Figure 6 shows the AFSW model over predicted the turbulent flame speed at equivalence ratio 0.43 whereas at equivalence ratio 0.5 and 0.56 predicted turbulent flame speed which is in good agreement with experiments. Turbulent flame speed is an important combustion parameter in study of combustion and the most difficult parameter to be able to predict by numerical models. The model showed no difference in flame lengths for all the equivalence ratios. Turbulent flame brush thickness is also in good agreement with experiments.

5. Conclusions

The study focused on the flow characterisation of the non-reacting and reacting flow field in the PSI high-pressure combustion test rig for methane/air lean mixtures. The operating conditions used are flow velocity of 40m/s, high pressure of 5bar and equivalence ratios of 0.43, 0.5 and 0.56. These conditions are mainly suitable for gas turbine power plants. An algebraic relation (AFSW model) for predicting the chemical burning rate in turbulent premixed flows has been implemented in the open source CFD code. The AFSW model is tested at various conditions to meet the experimental results through numerical simulations. The study suggested a need for fine-tuning of the model, hence a parametric variation study is carried out. This shows the pre-constant of AFSW model need to be tuned for each equivalence ratio which resulted in converged solutions. The tuned AFSW model predicted better turbulent flame speed and turbulent flame brush thickness values which are in good agreement with experiments. Further work is needed to achieve a universal relation valid for broad range of conditions.

References:

1. P. Griebel, E. Boschek et al., P. Jansohn, *Flame Stability and NOx Emission Improvements due to H_2 Enrichment of Turbulent, Lean Premixed, High-Pressure, Methane/Air Flames*, The Future of Gas turbine technology 3rd Intl. Conference 2006, Belgium.
2. Weller, H.G. *The development of a new flame area combustion model using conditional averaging.* Thermo-fluids section report TF/9307 Department of Mechanical Engineering Imperial College of Science Technology and Medicine, 1993.
3. SPR Muppala et al., *Development of an algebraic reaction rate closure for the numerical calculation of turbulent premixed methane, ethylene, and propane/air flames for pressure up to 1.0 MPa,* Combustion and Flame 140 (2005) 257-266.
4. N K Aluri, SPR Muppala et al., *Large Eddy Simulation of Lean Premixed Turbulent Flames of Three Different Combustion Configurations using a novel Reaction closure*, Flow Turbulence Combust (2008) 80:207-224.
5. N K Aluri and SPR Muppala et al., *Substantiating a fractal-based algebraic reaction closure of premixed turbulent combustion for high pressure and Lewis number effects*, Combustion and Flame 145(2006) 663-674.
6. Siva Muppala and Ramesh Badiger et al., *RANS Simulation of High-Pressure Premixed Turbulent Lean Flames in PSI Combustor,* Hyderabad Asia Pacific Conference – Dec-2010.

Numerical simulation of combustion instabilities under the alternating gravity conditions

Anastasia Krikunova*

Moscow Institute of Physics and Technology (State University), 141701, Dolgoprudny, Moscow Region, Russian Federation

Abstract. The work is devoted to the analysis of the methane-air conical flame behaviour under conditions of an alternating gravitational field. Numerical simulation based on the software package FlowVision, has shown the possibility of modeling the flame front instabilities during the transition from the normal gravitational conditions to zero gravity. The appearance of the flame front oscillations is demonstrated under the such conditions. Further studies will provide a complete picture of the behavior of the flame in an alternating gravitational field.

1 Introduction

Combustion processes experimental studies, both under the weightlessness conditions and under the increased gravity (overload) in centrifuges, Drop towers, ISS and parabolic flights, were carried out by various scientific centers, such as NASA, ESA, Roskosmos [1].In particular, the author of the current work carried out experiments on the vacuum-dynamic Bremen Drop tower facility [2 - 4].However, in the literature there is no data on the flame characteristics under transitional conditions (transition from the normal gravity to the microgravity and reverse), which is undoubtedly an essential issue. During the aircraft operating in the modes of such transitions, instabilities may occur, leading to the flame extinction/quenching, which may be critical for the engines stable operation. So the work is of interest, both from the point of view of fundamental research, and for applied activity.

2 Numericali simulation details

Numerical simulation of processes by the finite element method was carried out by using a FlowVision software package [5]. A detailed illustration of the nozzle geometry and the computational domain design is shown in the Figure. 1. The burner geometry was identical to the experimental conditions, that were carried out earlier [3]. A conical nozzle was used as the burner, providing a uniform flow of velocity at the outlet section. The geometric dimensions are indicated in Figure 1. Also the flame was stabilized by a ring, placed near the outlet edge of the nozzle. The numerical domain was chosen sufficiently large to avoid

* Corresponding author: utro-2007@mail.ru

the influence of the walls on the processes under the investigation, for example the interface between combustion products and ambient area should be free from the far field compression effects. A cylindrical domain with height of 175 mm and diameter of 70 mm was used. Normal temperature and pressure were used as the initial conditions; mass flow rate and fuel-oxygen ratio were settled in the inlet boundary condition. From the very beginning the computational domain was filled with air under the normal conditions. The non-uniform 3D grid model was used, it was built from the uniform grid with size 25x19x19 cells by using 6 order approximation in the vicinity of walls, ring-stabilization body and the flame front region. Finally grid was consisted of amount above 3.3 mln cells. A complete system of Navier-Stokes equations for a viscous compressible gas was simulated. The equations of state of the components corresponded to the model of an ideal gas. The k-ω SST turbulence model was chosen. To solve the system of linear algebraic equations, an implicit numerical scheme of the second order of accuracy without constraints was used. At the initial stage a time step corresponding to the Courant number of 100 was established to form a developed flow, and then it was decreased to CFL = 20. Ignition was carried out by raising the temperature at the nozzle edge vicinity for a short time up to 2000°K, while the CFL was increased to 50, after the formation of the flow of the combustible mixture. After the formation of a stable flame, the temperature source was removed and the time step was successively reduced up to 10^{-4} s. One-step global reaction model of methane combustion was used to simplify the modeling process and reduce the simulation time. The future work will be devoted to the detailed kinetic scheme introduction into the simulation code.

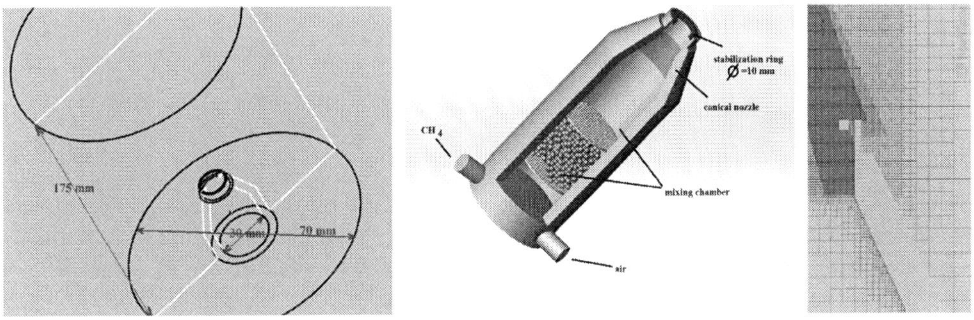

a b c

Fig. 1. Computational domain (a), burner (b), numerical domain (c).

3 Results and discussions

Figure 2 shows the distribution of temperatures and rates of combustion reactions ([kg m-3 s-1]) for normal gravity and zero gravity conditions (transition region). The position of the flame front can be determined from the position of the maximum temperature for the terrestrial gravity (Figures 2a, 2c, 2e), whereas under the transition to weightlessness, the temperature field is redistributed (Figures 2b, 2d, 2f). The position of the maximum temperatures region does not correspond to the region of maximum reaction rates and the position of the combustion flame front, since under reduced gravity conditions there are no convective flows that play a major role.

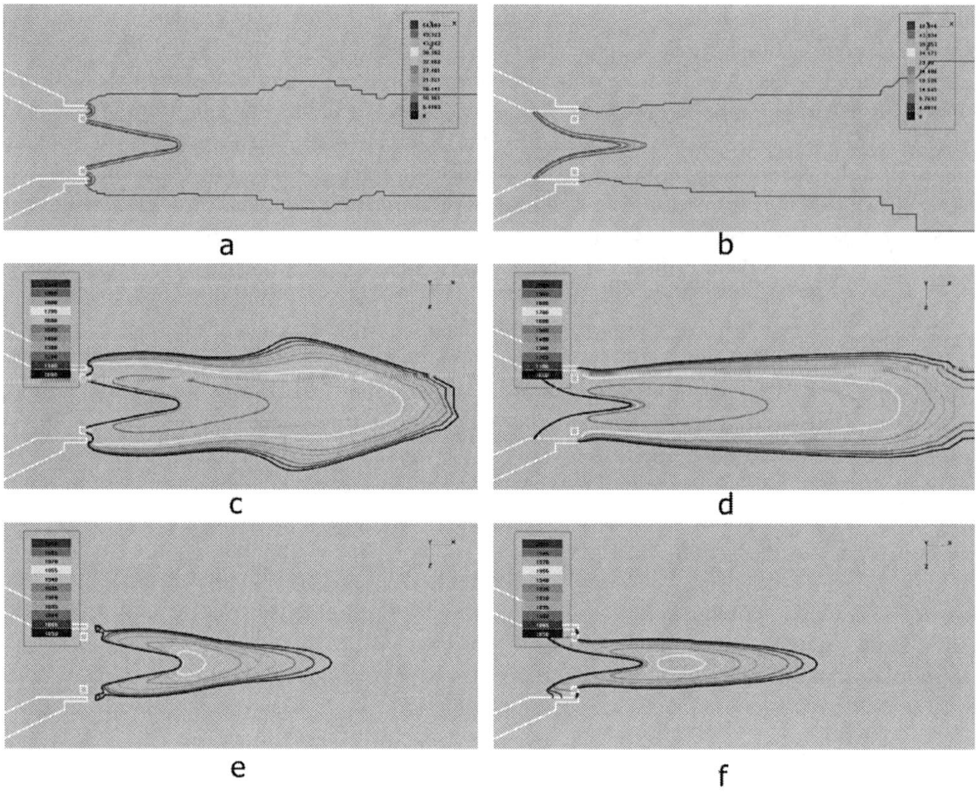

Fig. 2. Rates of combustion reactions under normal gravity conditions (a) and under the transition to the microgravity (b), temperature distribution for the normal gravity (c), (e) and transition conditions (d), (f).

Flame front instabilities occur during the transient conditions, which eventually decay and, as shown earlier [4], under the advanced microgravity conditions, the flame is subjected only to high-frequency oscillations. At the same time, the oscillations are so strong that the plume parts are separated from the main plume and moved downstream the flow. Figure 3 shows the temperature fields in successive time intervals, illustrating the formation, separation and removal of such a part. Such kind of flame front instabilities brings to that the flame front has the shape similar to the premixed flames under the acoustic perturbation. So it was shown the possibility to simulate such important flame characteristic as flame front instabilities under the alternating gravity conditions. The future work will be devoted to the analysis of the such kind instabilities and comparison with experimental results.

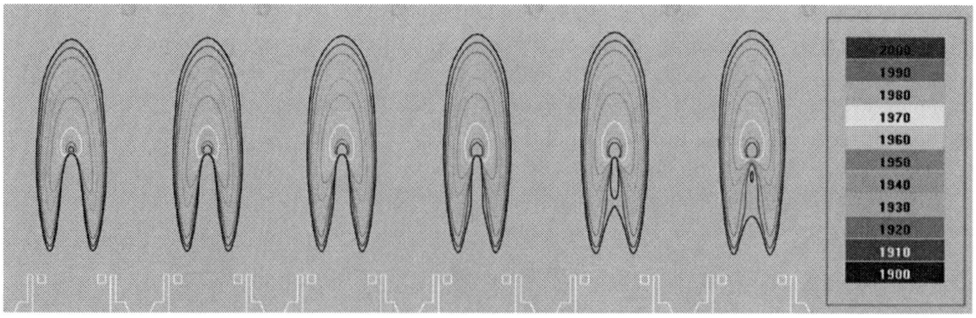

Fig. 3. Flame temperature field dynamics, time step - 0.001 s.

The work was supported by the Russian Science Foundation, Grant No. 18-79-00284.

References

1. H.D. Ross (Ed.), *Microgravity combustion: fire in free fall* (Academic press, 2001)
2. M. Dreyer, Microgravity Sci. Technol. **22,** 4 (2010)
3. A.I. Krikunova, E.E. Son, High. Temp. 1 (2018)
4. A.I. Krikunova, E.E. Son, Microgravity Sci. Technol. 1 (2018)
5. A.A. Aksenov, Computer Research and Modeling **9**, 5 (2017)

MATEC Web of Conferences **209**, 00006 (2018) https://doi.org/10.1051/matecconf/201820900006
ComPhysChem'18

Measurements of rate constants of O_2(b) quenching by CH_4, NO, N_2O at temperatures 300-800 K

G.I. Tolstov[1,2*], *M.V.* Zagidullin[1,2], *N.A.* Khvatov[1,2], I.A. Medvedkov[1,2], *A.M.* Mebel[1,3], *M.C.* Heaven[1,4] and *V.N.* Azyazov[1,2]

[1]Samara University, Samara, 443086, Russia
[2]Lebedev Physical Institute of RAS, Samara, 443011, Russia
[3]Florida International University, Miami, 33199, USA
[4]Emory University, Atlanta, Georgia 30322, USA

Abstract. Electronically excited oxygen has an important place in the kinetic schemes of the processes taking place in the atmosphere, in the active medium of an oxygen-iodine laser, and in plasma-assisted combustion[1]. Over the past decades, a large amount of data on the rate constants of quenching O_2(b) on a large number of collision partners has been accumulated. However, they mostly refer to the results of measurements at room temperature. In this paper, rate constants for the quenching of O_2(b) by collisions with N_2O, NO, and CH_4 have been determined in the temperature range from 297 to 800 K, by the laser-induced fluorescence method. O_2(b) was excited by pulses from a tunable dye laser, and the deactivation kinetics were followed via observing the temporal behavior of the $b^1\Sigma_g^+ \rightarrow X^3\Sigma_g^-$ fluorescence. From the analysis of experimental results, the following temperature dependencies of the quenching rate constants by these gases were obtained, and could be represented by the expressions: $k_{NO}=(1.77\pm0.2)\times10^{-24}\times T^{3.5}exp(1138\pm37/T)$; $k_{N2O}=(2.63\pm0.14)\times10^{-16}\times T^{1.5}\times exp(590\pm26/T)$ and $k_{CH4}=(3.54\pm0.4)\times10^{-18}\times T^{1.5}\times exp(-220\pm24/T)$ cm^3s^{-1}. All of the rate constants measured at room temperature were found to be in good agreement with previously reported values.

1 Introduction

Gas phase processes with oxygen molecules in second electronically excited state $b^1\Sigma_g^+$ has been studied extensively in last decades, due to the importance of these processes in the terrestrial atmosphere [1-3], oxygen-containing gas discharges [4-6] and combustion [7-10]. Plasma-chemicals processes in the air-fuel mixtures result in formation of $O_2(b^1\Sigma_g^+)$ molecules that may activates chain reactions in the combustion zones [11]. Analysis of

* Corresponding author: outloot@yandex.ru

reaction kinetics involving $O_2(b^1\Sigma_g^+)$ is made difficult, because there is little published data concerning the deactivation kinetics at temperatures above 350 K.
Studies of the $O_2(b^1\Sigma_g^+)$ deactivation at room temperature by molecules relevant to combustion are represented by many works: CH_4 [12-17], NO [18], N_2O [12,13,18-20]. So, the aim of this work was to determine the rate constants for removal of $O_2(b^1\Sigma_g^+)$ by the CH_4, NO, and N_2O molecules at temperatures up to ~ 800K using time-resolved emission from the O_2 $b^1\Sigma_g^+$ - $X^3\Sigma_g$ transition.

2 Experiment

The measurements were made by observing the decay profiles of emission intensity from excited oxygen $O_2(b^1\Sigma_g^+)$ on the addition to the oxygen flow of the quencher, using the excitation of $b^1\Sigma$ state by pulsed dye laser. An oxygen-quencher gas mixture flowed through the fluorescence cell (FC), which consisted of a cylindrical quartz tube 40 cm long, with an internal diameter of 15 mm, the ends were sealed by quartz windows, mounted at an angle. The oxygen-containing mixture was prepared right before the entrance to the FC. The mixture was heated via the heat transfer from the warmed walls of the quartz tube. The temperature of the walls was maintained by resistive heating of the nichrome tape wound on it and the corresponding thermal insulation. To observe emission from $O_2(b^1\Sigma_g^+)$ a rectangular window, 2 cm long and 1 cm high was located in the center of FC. This aperture was free of insulating materials and heating tape. The gases used had the following minimum purities stated by the manufacturers: O_2(>99.99995), CH_4 (>99.99%), N_2O (>99.8%, H_2O<0.025%), NO, (>99.9%) were used in the experiments. The gas temperature in the LIF region was measured by a thermocouple.

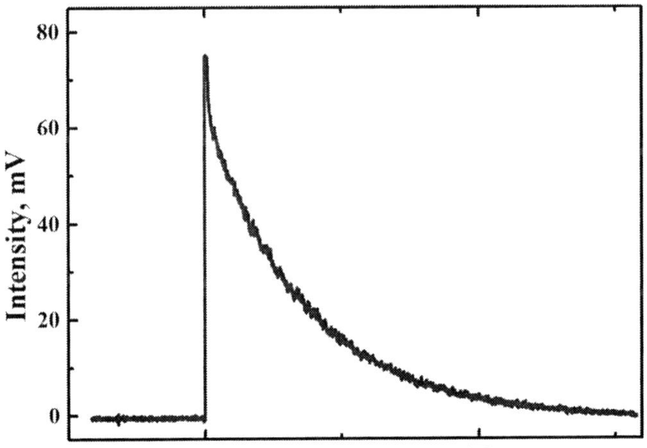

Fig. 1. Temporal profiles of the $O_2(b)/CH_4$ mixture recorded for oxygen partial pressure P_{O2} = 657.3 Torr, and CH_4 number density n_{CH4} = 4.11×10^{-16} cm^{-3} at gas temperature T=300K

The dye laser beam (Sirah Precision Scan, PSCAN-D-18-EG) was directed along the central axis of the FC. This laser pulse excited oxygen molecule into ($b^1\Sigma_g^+$, υ=1) state. In excess $O_2(X)$, the $O_2(b^1\Sigma_g^+$, υ=1) was rapidly transferred to $O_2(b^1\Sigma_g^+$, υ=0) by efficient E–E and V–V energy-exchange processes[21]. Oxygen $^1\Sigma_g^+$, υ=0 → $^3\Sigma_g^-$, υ=0 emission was observed through the window in FC, along an axis that was perpendicular to the dye laser beam. The light was collected and focused on the entrance slit of a monochromator (MDR-206), transmitting wavelengths in the 762 ± 8 nm range. Additional long-pass filters were used to suppress the scattered radiation at 690 nm. Time-resolved LIF signals were recorded using a PMT (Hamamatsu R636-10). To measure time resolved fluorescence decay curves, the PMT output electric signals stored in memory of digital oscilloscope (LeCroy Wavesurfer-3054R).

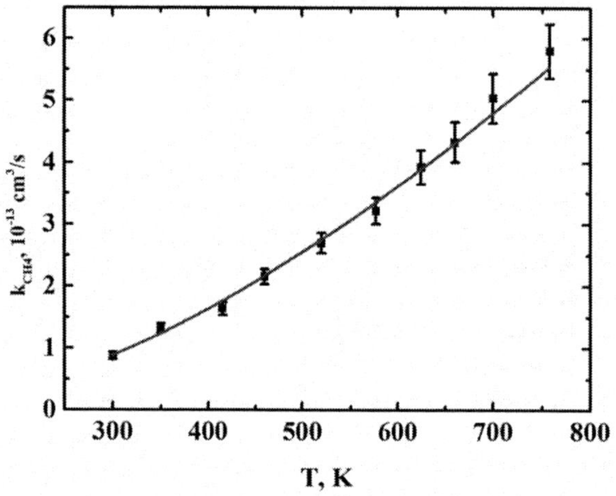

Fig. 2. Rate constant k_{CH4} as a function of temperature. Solid curve is the plot of $3.54\times10^{-18}T^{1.5}exp(-220/T)$.

3 Results

The PMT response temporal profile shown on Figure 1 have bi-exponential behavior due to the instrumental decay caused by short (10 ns) scattering emission and the decay of $^1\Sigma_g^+ \rightarrow {}^3\Sigma_g^-$ emission. The decay of these profile can be fitted by two exponents: $A_0 + A_r \exp(-K_r t) + A exp(-Kt)$, where $(1/K_r)$ is the instrumental response time of the order of 2μs. The decay rate of $O_2(b^1\Sigma_g^+)$ removal K in the processes

$$O_2(b^1\Sigma_g^+) + O_2 \rightarrow 2O_2 \quad (1)$$

$$O_2(b^1\Sigma_g^+) + M \rightarrow O_2 + M. \quad (2)$$

is given by expression

$$K = k_{O2} n_{O2} + k_M n_M, \quad (3)$$

where n_{O2} and n_M are the number densities of oxygen O_2 and an admixed gas M, k_{O2} and k_M are rate constants of processes (1) and (2) respectively. Application of the ideal gas approximation yields the expression [22]

$$k_m - K\left(\frac{k_B T}{P}\right)\frac{G}{G_{O2}},$$

where

$$k_m = k_{O2} + \frac{G_M}{G_{O2}} k_M \quad (4)$$

is the combined rate coefficient. Here P is the gas pressure, T is the gas temperature, k_B is the Boltzmann constant, G is the total flow rate, G_M is the flow rate of the admixed gas M.

The plot of k_m versus G_M/G_{O2} for flowing gas is the analogues to the Stern-Folmer plot of K versus number density of M for static condition. The slope of linear fit according to equation (4) gives rate constant k_M. Then the set of data of k_M vs T was fitted according to equation

$$k_M(T)=AT^{0.5n}\exp(-E/T) \quad (5)$$

with integer value for n was used to obtain analytical temperature dependence of rate constants.

We observed that the temperature of the gas mixture in the LIF observation region, directly opposite the center of the fluorescence observation window, was somewhat lower than the upstream or downstream temperatures. The temperature gradient was present because the higher heat transfer through the window in the wall of the quartz tube. Temperature variations were taken into account in estimating the measurement errors.

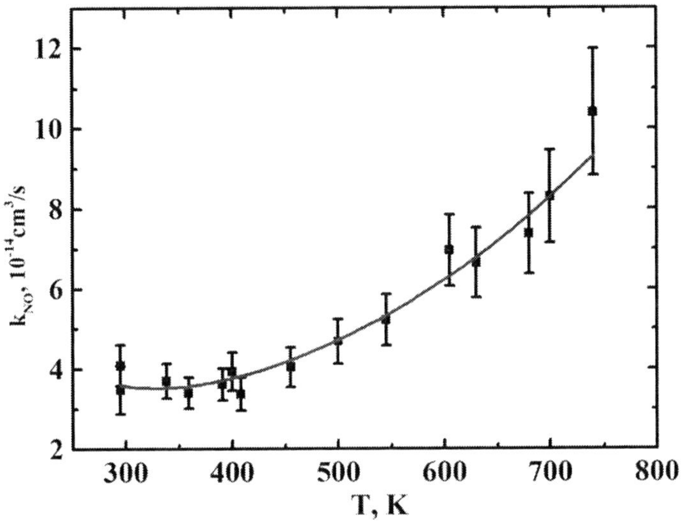

Fig. 3. Rate constant k_{NO} as a function of temperature. Solid curve is the plot of $1.77\times10^{-24}\times T^{3.5}exp(1138/T)$ fitting experimental data (symbols).

The results of measurements (symbols) and plots of equations for $k_M(T)$ for removal species are shown in Figures 2-4.

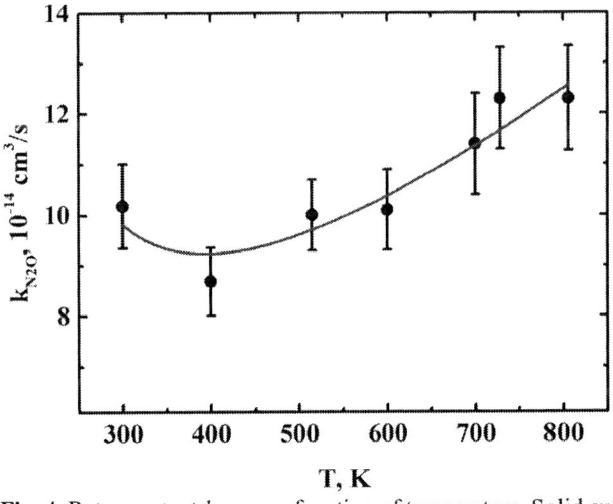

Fig. 4. Rate constant k_{N2O} as a function of temperature. Solid curve is the plot of $2.63\times10^{-18}T^{1.5}exp(590/T)$ fitting experimental data (symbols).

4 Conclusion

The measured values of the rate constants for room temperature are in good agreement with the results of other studies. For each of the collisional partners M the fitting expression of the type (5) was derived in the temperature range 297-800 K. The measured values of the rate constants and fitting expressions for $k_M(T)$ can be used in modeling of the gas kinetics processes involved $O_2(b^1\Sigma_g^+)$ molecules in the temperature range 297-800 K.

This work was supported by the Ministry of Education and Science of the Russian Federation under the Grant No. 14.Y26.31.0020 to Samara University.

References

1. T.G. Slanger, R.A Copeland, Chem. Rev. **103** (2003).
2. A.S. Kirillov, Chem. Phys., **410** (2013).
3. V.A. Krasnopolsky, *Planetary and Space Science*, **59**, (2011).
4. G. Nayak, J.S. Sousa, P.J. Bruggeman, *Journal of Physics D: Applied Physics*, **50**, (2017).
5. T. Wegner, C. Küllig, J. Meichsner, *Plasma Sources Science and Technology*, **26**, (2017).
6. Y.A. Mankelevich, E.N. Voronina, A.Y. Poroykov, T.V. Rakhimova, D.G. Voloshin, A.A. Chukalovsky, *Plasma Physics Reports*, **42**, (2016).
7. A.M. Starik, A.V. Pelevkin, N.S. Titova, *Combustion and Flame*, **176**, (2017).
8. A.A. Konnov, *Combustion and Flame*, **162**, (2015).
9. L.V. Bezgin, V.I. Kopchenov, A.M. Starik, N.S. Titova, *Combustion, Explosion, and Shock Waves*, **53**, (2017).
10. A.V. Pelevkin, B.I. Loukhovitski, A.S. Sharipov, *The Journal of Physical Chemistry A*, **121**, (2017).
11. A. Starik, A. Sharipov, *Physical Chemistry Chemical Physics*, **13**, (2011).
12. S.V. Filseth, A. Zia, K.H. Welge, J.Chem.Phys., **52**. (1970).
13. K.H. Becker, W. Groth, U. Schurath, Chem.Phys.Lett., **8**, (1971).
14. J. Wildt, G. Bednarek, E.H. Fink, R.P. Wayne, *Chem. Phys.*, **122**, (1988).
15. K. Kohse-Höinghaus, F. Stuhl, J. Chem. Phys., , **72**, (1980).
16. J.A. Davidson, E.A. Ogryzlo, Can. J. Chem., **52**, (1970).
17. E.J. Dunlea, R.K. Talukdar, A.R. Ravishankara, J. Phys. Chem. A, **109**, (2005).
18. R. Boodaghians, P. Borrell, P. Borrell,. Chem.Phys.Lett., **97**, (1983).
19. P.M. Borrell, P. Borrell, K.R. Grant, The Journal of Chemical Physics **78**, (1983).
20. E.J. Dunlea, R.K. Talukdar, A.R. Ravishankara, Zeitschrift für Physikalische Chemie., **224**, (2010).
21. H.I. Bloemink, R.A. Copeland, T.G. Slanger. J. Chem. Phys., **109**, (1998).
22. M.V. Zagidullin, N.A. Khvatov, I.A. Medvedkov, G.I. Tolstov, A.M. Mebel, M.C. Heaven, V.N. Azyazov, *J. Phys. Chem. A*, **121**, (2017).

Mathematical modelling of burning surface in parallel flow of oxidant

Veronika Tyurenkova[1,2,*], *Nickolay* Smirnov[1,2], and *Mariya* Smirnova[1,2]

[1]Federal Science Center Scientific Research Institute for System Studies of Russian Academy of Sciences, 117218 Moscow, Russia
[2]Moscow M.V. Lomonosov State University, 119991, Moscow, Russia

Abstract. The paper presents the problem of condensed material surface burning in a flow of oxidant within the framework of the assumptions of the laminar boundary-layer theory. It is used assumption of fuel gasification and gas phase chemical reacting in a diffusion flame. The regression rate of the material surface in the turbulent and laminar flow regimes is studied. The zones correspond to kinetic and diffusion regime is defined.

1 Introduction

The condensed fuel burning in an oxidant gas flow occurs in investigations of various processes, for example: burning of fuel in hybrid engines; interaction of hypersonic flying vehicles with the atmosphere; exo- and endothermic reactions in chemical engineering; spreading of flame over the fuel surface, the origination of explosions and the propagation of detonation in unmixed two-phase systems of the gas-film type.

The present work investigates the burning of the flat surface of fuel in an oxidant flow in the absence of an axial pressure gradient. The problem is solved within the framework of the assumptions of the laminar boundary-layer theory. Near the fuel surface in the gas, the boundary layer develops in the interior of which there proceed chemical reactions between the fuel vapours and the oxidant. Since these are exothermic reactions, energy is given up by the boundary layer in the gas. The heat from the heated gas transfers to the fuel layer which, having evaporated, gains access, as a result of diffusion, the boundary layer, where it enters into reaction with the oxidant. As a result of continuous inflow of heat from the gas phase, the fuel becomes warmed up and a temperature boundary layer develops near the phase interface.

2 Mathematical problem statement

The main aspect of the problem statement is the model for gasification of solid fuel surface. As for now, there exist two different approaches, which represent the two limiting case: quasi-equilibrium gasification and absolutely non-equilibrium gasification.

Let the gasification process be described by Hertz-Knudsen formula

* Corresponding author: tyurenkova.v.v@yandex.ru

$$\dot{m}_N = k(p_N^* - p_N),\; k = \delta\sqrt{\frac{m_N}{2\pi RT_w}} = \delta_o\sqrt{\frac{m_N}{2\pi RT_w}}e^{-\frac{E_\delta}{RT_w}}, \tag{1}$$

where p_N is partial pressure of fuel vapor near the phase interface, T_w - temperature of the interface, p_N^* - phase equilibrium pressure for $T = T_w$, m_N - molar mass of fuel, R - universal gas constant, δ - accommodation coefficient, δ_0 - maximal allowable value for accommodation coefficient, E_δ - activation energy for gasification. The limiting cases of for formula (1) are the following:

(1) For relatively high partial pressure of the fuel vapor in the flow an equilibrium state at the interface can be achieved ($p_N^*(T_w) = p_N$) under which the numbers of gasifying and condensing molecules are equal. However, in the presence of heat flux phase transition rate can be determined then based on the energy balance, which is not taken into account by classical Hertz-Knudsen model.

(2) For relatively low partial pressure of fuel vapor ($p_N << p_N^*(T_w)$) formula (1) can be presented in the form

$$\dot{m}_N = \delta_o\sqrt{\frac{m_N}{2\pi RT_w}}p_N^* e^{-\frac{E_\delta}{RT_w}} \tag{2}$$

which is the gasification (or pyrolysis) formula accepted for many solid fuels.

(3) For the intermediate case the partial pressure being non-zero, and, yet, rather far from its equilibrium value, the formula (1) should be used along with the energy flux conservation condition to determine the unknown values of $\dot{m}_N$ and T_w. This generalized non-equilibrium model was developed and applied for droplets combustion in [1, 2].

The solution of the problem will be based on the concept of diffusion flame and will use analytical research methodology developed in [1-2]. Contrary to symmetric problems solved in [1-2] the present problem statement will differ greatly by incorporating the flow dynamics in the frame of boundary layer approximation, and accounting for fuel heating dynamics on moving from the flame tip.

Let the rectangular coordinate systems (x, y) and $(\tilde{x}, \tilde{y})$ be chosen so that the axis x would be coincident with the phase interface and would be aligned with the oxidant flow, while the axis y and $\tilde{y}$ in the gas and liquid, respectively, would be aligned against the interface. The coordinate origin coincides with the start of the boundary layer.

Then, the equations of a boundary layer in a gas, provided that the Prandtl and Schmidt numbers are equal to unity, are:

$$\frac{\partial \rho u}{\partial x} + \frac{\partial \rho v}{\partial y} = 0,\; \rho u\frac{\partial u}{\partial x} + \rho v\frac{\partial u}{\partial y} = \frac{\partial}{\partial y}\mu\frac{\partial u}{\partial y} \tag{3}$$

$$\rho u\frac{\partial Y_i}{\partial x} + \rho v\frac{\partial Y_i}{\partial y} = \frac{\partial}{\partial y}\mu\frac{\partial Y_i}{\partial y} + \omega_i; i = 1,\dots,N,\; \rho u\frac{\partial H}{\partial x} + \rho v\frac{\partial H}{\partial y} = \frac{\partial}{\partial y}\mu\frac{\partial H}{\partial y} \quad p = \rho RT\sum_{i=1}^{N} Y_i / m_i$$

where $H = \sum_{i=1}^{N} Y_i(c_{pi}T + h_i^o) + \frac{u^2}{2}$ is the total specific enthalpy of the mixture and, a $h_i = c_{pi}T + h_i^o$ is the i-th component specific enthalpy. In the case of solid fuel the equations for a boundary layer at constant heat capacity c_f, viscosity μ_f and thermal conductivity λ_f reduce to energy equation:

$$\tilde{v}_f \frac{\partial \tilde{T}}{\partial \tilde{y}} = \frac{\lambda}{\rho c} \frac{\partial^2 \tilde{T}}{\partial \tilde{y}^2} \tag{4}$$

where $v_f = v_f(x)$ is the velocity of vertical movement of fuel, deriving from evaporation, because coordinate system in the vertical direction is moving with the phase interface.

The conditions on the outer edge of a boundary layer in a gas ($y \to +\infty$): $T = T_e, p = p_e, u = u_e, Y_1 = Y_{1e}, Y_{1e} + Y_{2e} = 1$. The conditions on the edge of a boundary layer in a fuel ($\tilde{y} \to +\infty$): $\tilde{T} = T_\infty, \tilde{u} = u_\infty = 0$. The balance relationships at the phase interface ($y = 0; x > 0$) are following: total mass flux conservation condition $(\rho v)_w = -(\rho\tilde{v})_{fw}$, no-slip condition $u = \tilde{u} = 0$, condition of tangential stresses continuity $\tau = -\tau_{fw}, \left(\tau = -\mu \frac{\partial Y_i}{\partial y} \right)$.

Condition of variation in the mass flow of each of the components in the absence of heterogeneous reactions $(\rho v)_w (Y_{iw} - Y_{iwf}) - (\rho D)_w \left(\frac{\partial Y_i}{\partial y} \right)_w = 0$. Total energy flux conservation condition under assumptions that: the fuel is one-component ($Y_{Nfw} = 1$), the heterogeneous reactions are absent ($g_{iw} = 0$), the squares of the velocity components on the axis y (v_{fw}^2, v_w^2) may be neglected, has the following form $\lambda_f \left(\frac{\partial \tilde{T}}{\partial \tilde{y}} \right)_{fw} = -\lambda \left(\frac{\partial T}{\partial y} \right)_w + (\rho v)_w h_L$.

Define the interface temperature T_w as the temperature of equilibrium evaporation of fuel at a pressure p_{Nw}, which is equal to the partial vapor pressure of the vaporized fuel at the interface $p_{Nw} = Y_{Nw} p_e \frac{m_w}{m_N}$. The molecular weight of the mixture m_w is determined by the following equation $\frac{1}{m_w} = \sum_{i=1}^{N} \frac{Y_{iw}}{m_i}$. The pressure p_{Nw} could be defined from the Clausius-Clapeyron equation $\frac{p_e}{p_{NW}} = \left(\frac{T_{ke}}{T_w} \right)^{\Delta c_p m_N / R} \exp\left[\frac{\Delta c_p m_N T_{kp}}{R} \left(\frac{1}{T_{ke}} - \frac{1}{T_w} \right) \right]$, where T_{ke} - is the phase transition temperature at the pressure p_e, T_{kp} - critical temperature.

3 Self-similar problem solution

Introduce new independent variables $z = \int_o^y \rho dy, \varsigma = \rho\mu x$, the dependent variables $w = \frac{1}{\rho\mu}(\rho v + u \frac{\partial z}{\partial x})$ and steam functions Ψ so that $u = \frac{\partial \Psi}{\partial z}$; $w = -\frac{\partial \Psi}{\partial \varsigma}$.

Then, the first equations of system (3) are satisfied identically, while the remaining equations, after the transition to the similarity variables $\eta = z\sqrt{\frac{u_e}{2\varsigma}}$; $\Psi = \sqrt{2\varsigma u_e} g(\eta)$ are reduced to the ordinary differential equations

$$g'''(\eta) + g''(\eta) g(\eta) = 0 \tag{5}$$

with boundary conditions

$$g'(0)=0, g'(\infty)=1, g(0)=-Bg''(0), \tag{6}$$

where $B=\frac{(\rho v)_w u_e}{\tau_w}$ is the dimensionless parameter of mass transfer.

The equation (4) can be integrated: $-\left(\lambda\frac{\partial\tilde{T}}{\partial\tilde{y}}\right)_{fW}=(\rho\tilde{v})_f c_f(T_\infty-T_W)=(\rho v)_w c_f(T_w-T_\infty)$.

Thus the equation (5) and boundary conditions (6) make it possible to solve the problem. The surface temperature T_w and the mass transfer parameter B are determined by the equations $B=Y_{1e}\Phi_N+\sum_{i=2}^{N-1}\frac{m_N}{m_i}(Y_{ie}-Y_{1e}\Phi_i)(\frac{p_e}{p_{NW}}-1)^{-1}$ and

$$B[c_f(T_w-T_\infty)+h_L]=\hat{c}_{pe}(T_e-T_w)+\frac{u_e^2}{2}+Y_{1e}(\Phi_N\Delta H+\Delta C_p T_w).$$

4 Regression rate

The regression rate is an important parameter for the hybrid rocket motor, and many efforts have been focused to study the regression rate. The required rate can be determined by the Arrhenius law $\dot{r}=A\exp(-E_a/RT_w)$, where A, E_a, R are constants, T_w is the temperature at the interface, as it is done in many papers. Considering the results derived in the previous chapter, we obtain an analytical solution for regression rate other than described. The regression rate can be determined by the Arrhenius law $\dot{r}=A\exp(-E_a/RT_w)$, where the parameters A, E_a for HTPB is $A=0.01104$, $E_a=20523.8$ for $T_w>722K$ and $A=3.96480$, $E_a=55803$ for $T_w<722K$.

Fig. 1 shows the regression rate profiles, which were obtained, first, based on self–similar solution of quasi-equilibrium problem (curves 1 – 3) $\tilde{v}_f^{lam}=1/\rho_f(\sqrt{(u_e\rho\mu/2)}g(0)/\sqrt{x})$ and, second, based strictly non-equilibrium model (Arrhenius law, curve 4). It is seen from Fig. 1, that the regression rate, determined by the Arrhenius law, is constant, which is valid for very high diffusion fluxes, when kinetics of gasification becomes the limiting factor. In burning solid fuel surface kinetic regime is realized not for the whole specimen, but only near the tip of boundary layer. At longer distances from the tip diffusion fluxes decrease thus becoming the limiting factor. Since kinetic regime (2) presents the fastest gasification rate, the self-similar solution near the tip surpassing the kinetic regression rate cannot be valid, as well as kinetic regime is not valid at longer distances from the tip, where it is limited by diffusion fluxes.

Depending on the flow velocity the flow may be laminar and turbulent in the boundary layer. As a rule, there is a combined flow in the boundary layer characterized by transition from laminar to turbulent regimes. The position of the transition zone can be approximately determined by the critical value of the local Reynolds number $\mathrm{Re}=\rho u_e x/\mu$. To find an analytical solution for the regression rate at a turbulent flow mode, we determine the tangential stress in the absence of mass transfer parameter (B=0) $\tau_{wo}=0.0296\cdot\rho u_e^2\cdot\mathrm{Re}^{-1/5}$. A link between tangential stresses in the presence of mass transfer τ_w and in the absence of one τ_{wo} can be obtained for small values of mass transfer parameter B $\tau_w/\tau_{wo}=\ln(1+B)/B$. Then the regression rate for turbulence flow regime $(\rho\tilde{v})_f=(\tau_w B)/u_e=0.0296\cdot\rho u_e\ln(1+B)\cdot\mathrm{Re}^{-1/5}$. Fig. 2 shows the comparison of the analytical solution with numerical solutions for regression rate of HTPB fuel. The numerical calculations are based on two methods of modeling the mass regression rate. In this problem, numerical results demonstrate a good coincidence with exact analytical solutions as well as with experimental data.

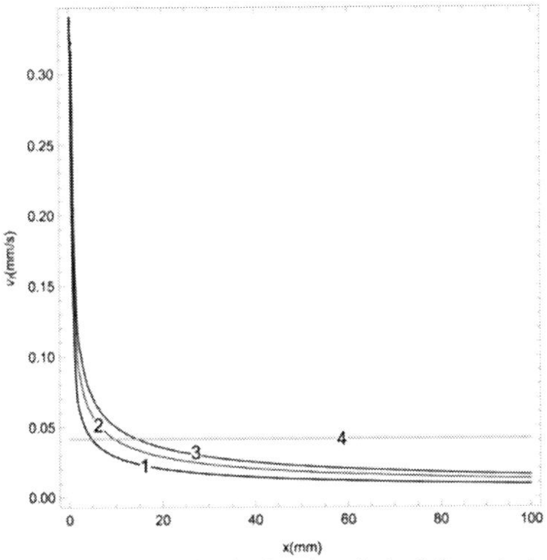

Fig.1. The profiles of regression rate, built on the basis of analytical formula (curve 1-3) and the Arrhenius law (curve 4) (curve 1 - $u_e = 5$ *m/s*, curve 2 - $u_e = 10$ *m/s*, curve 3 - $u_e = 15$ *m/s*).

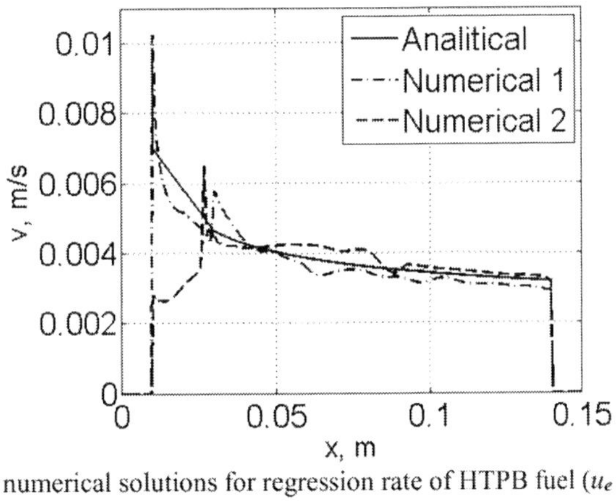

Fig.2. Analytical and numerical solutions for regression rate of HTPB fuel ($u_e = 1067$ *m/s*).

The authors gratefully acknowledge the financial support from the Russian Foundation for Basic Research (Grant № 17-07-00549).

References

1. V.V.Tyurenkova. Acta Astronaut. **75**, 78–84, (2012).
2. V.V.Tyurenkova, N.N. Smirnov, V.M. Guendugov. Acta Astronaut. **83**, 208–215, (2013).

Quantum chemical study of the mechanism of oxidation of $C_{15}H_9$ by atomic oxygen

Galiya Galimova[1], *Valeriy* Azyazov[1,2], *Alexander* Mebel[1,3*]

[1]Samara University, Samara, 443086, Russia
[2]Lebedev Physical Institute, Samara, 443011, Russia
[3]Department of Chemistry and Biochemistry, Florida International University, Miami, FL 33199, USA

Abstract. An important environmental problem related to the use of fossil fuels is the formation of soot during combustion. Mechanisms of soot oxidation, which alleviates its emission into the environment, are not fully understood. The reaction of O with a radical $C_{15}H_9$ may play an important role in combustion. In this article, the $C_{15}H_9$ molecule was chosen as a model of soot surface. The paper discusses various pathways resulting from the $C_{15}H_9$ + O reaction. Relative energies, frequencies and optimal geometries of the reactants, products, intermediates and transition states of the $C_{15}H_9$ + O reaction have been calculated using the quantum-chemical Gaussian and Molpro program packages. The reaction pathways leading to carbon monoxide CO elimination have been found and dscussed.

1 Introduction

One of the major harmful emissions occuring during combustion of hydrocarbon fuels is the emission of soot. Soot believed to be formed as aggregates of polycyclic aromatic hydrocarbons (PAH) [1]. The mechanisms of soot formation and destruction in hydrocarbon flames are extremely complex.

It is established that O_2, OH and atomic O are the major oxidizers of soot. Present kinetic combustion models, as a rule, describe soot oxidation as a result of collision between O_2, OH and O and the surface of soot particles. The reactions lead to degradation of soot when it eliminates a CO molecule.

The $C_{15}H_9$ molecule can serve as a model for a soot surface radical site produced by an H atom abstruction/elimination from barely reactive closed-shell soot surface. The purpose of the present study is to investigate the reaction pathways of $C_{15}H_9$ + O, to find the optimal geometry of the structures obtained, and to calculate vibrational frequencies and energies for all the considered reactants, products, intermediates and transition states, which can be utilized in future calculations of reaction rate constants and kinetic modeling of soot oxidation.

* Corresponding author: mebela@fiu.edu

2 Theoretical methods

Ab initio electronic structure calculations have been applied to investigate possible products of the $C_{15}H_9$ + O reactions and to find the reaction paths that are energetically feasible for this reaction. The geometries of various species (reactants, products, intermediates, and transition states) involved in the $C_{15}H_9$ + O reactions were optimized using the hybrid density functional B3LYP method with the 6-311G** basis set. The same B3LYP/6-311G** approach was used to compute vibrational frequencies and zero-point energy (ZPE) corrections.

Next, the G3(MP2,CC)//B3LYP version [2] of the original Gaussian 3 (G3) scheme for high-level single-point calculations was applied to obtain more accurate energies of all species. The B3LYP and MP2 calculations were performed using the Gaussian 09 program package [3], whereas the MOLPRO 2010 program package [4] was employed for RHF-RCCSD(T) calculations.

3 Results and discussion

We consider the reaction mechanism of the O atom with a five-member ring radical embedded in a sheet of six-member rings, like at the edge of a graphene sheet or a large PAH molecule. The oxidation process with atomic oxygen is emulated by the O reactions with $C_{15}H_9$ radicals made from three six-member and one five-member rings. Figures 1 and 2 show two different possible modes of embedding represented by two $C_{15}H_9$ isomers, (A) and (B). Isomer (A) has the embedded five-member ring sharing two edges (C-C bonds) with other six-member rings, whereas in (B) three C-C bonds in the five-member ring are shared with the six-member rings and the embedding is deeper. The computed potential energy diagrams for the reactions of these isomers with atomic oxygen are also shown in Figs. 1 and 2. While the general mechanistic features of the $C_{15}H_9(A)$ + O and the $C_{15}H_9(B)$ + O reactions are similar, the details and the energetics differ substantially.

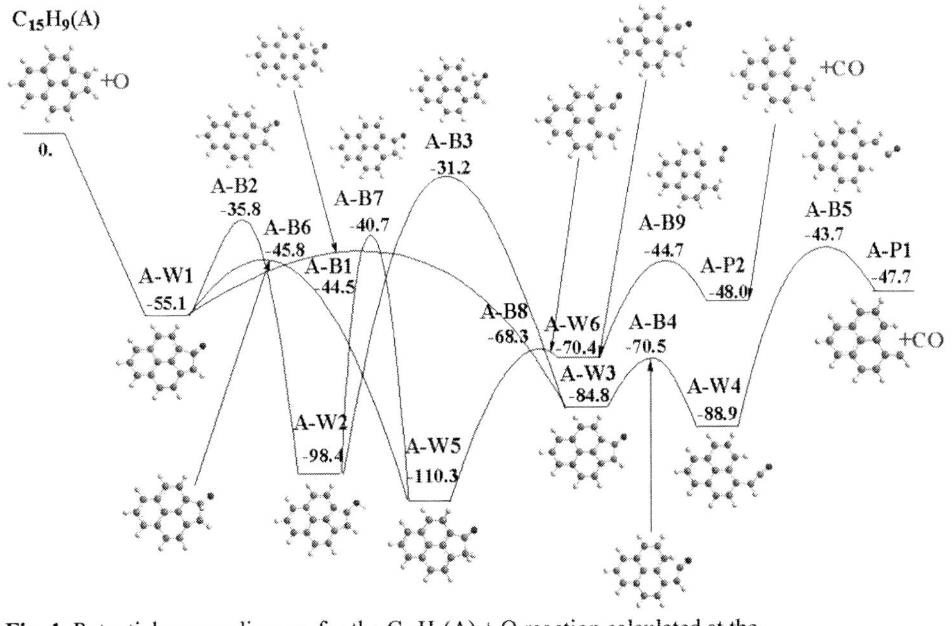

Fig. 1. Potential energy diagram for the $C_{15}H_9(A)$ + O reaction calculated at the G3(MP2,CC)//B3LYP/6-311G(d,p) level of theory. All relative energies are given in kcal/mol with respect to the initial reactants.

Figure 1 shows the $C_{15}H_9$(A) + O reaction, where barrierless O addition to the five-member ring in the entrance channel is 55.1 kcal/mol exothermic and forms the intermediate A_W1. A_W1 can evolve to $C_{14}H_9$ + CO (A_P1) by isomerization and decarbonylation via the following pathways: A_W1 → A_B1 (A_B2 → A_W2 → A_B3) → A_W3 → A_B4 → A_W4 → A_B5 → A_P1. These channels involve either direct formation of A_W3 by H migration from the C atom attacked by O to the *ortho* C atom common for the five- and six-member rings or via a two-step process with an H shift from carbon to oxygen forming A_W2 followed by a second H shift from O to *ortho* C. A_W3 then features the five-member ring opening forming A_W4; the latter eliminates the CO group and produces $C_{14}H_9$ + CO (A_P1) residing 47.7 kcal/mol below the initial reactants. The critical barriers on these pathways appeared to be A_B3 or A_B5, which lie 31.2 and 43.7 kcal/mol lower in energy than the $C_{15}H_9$(A) + O reactants, respectively.

The alternative reaction channel is A_W1 → A_B6 (A_B2 → A_W2 → A_B7) → A_W5 → A_B8 → A_W6 → A_B9 → A_P2. The pathways in this channel involve either direct formation of A_W5 by H migration from the attacked carbon to the *ortho* C atom exclusively belonging to the five-member ring or via a two-step process through the intermediate A_W2. A_W5 then features opening of the ring and formation of A_W6; the latter eliminates the CO group and produces $C_{14}H_9$ + CO (A_P2), 48.7 kcal/mol below the reactants. The critical barriers on these pathways are found at A_B2 and A_B7 residing 35.8 and 40.7 kcal/mol below the initial reactants. The difference in the structures of the A_P1 and A_P2 products is that in the former the radical is located on the side chain, while in the latter the radical is on the six-member ring. In both mechanisms, the five-member ring of $C_{15}H_9$ is annihilated in the reaction of $C_{15}H_9$(A) with atomic oxygen producing radicals consisting of only three aromatic six-member rings plus CO. Calculations of rate constants and product branching ratios are planned in the future to quantify the competition between the two ways of decarbonylation.

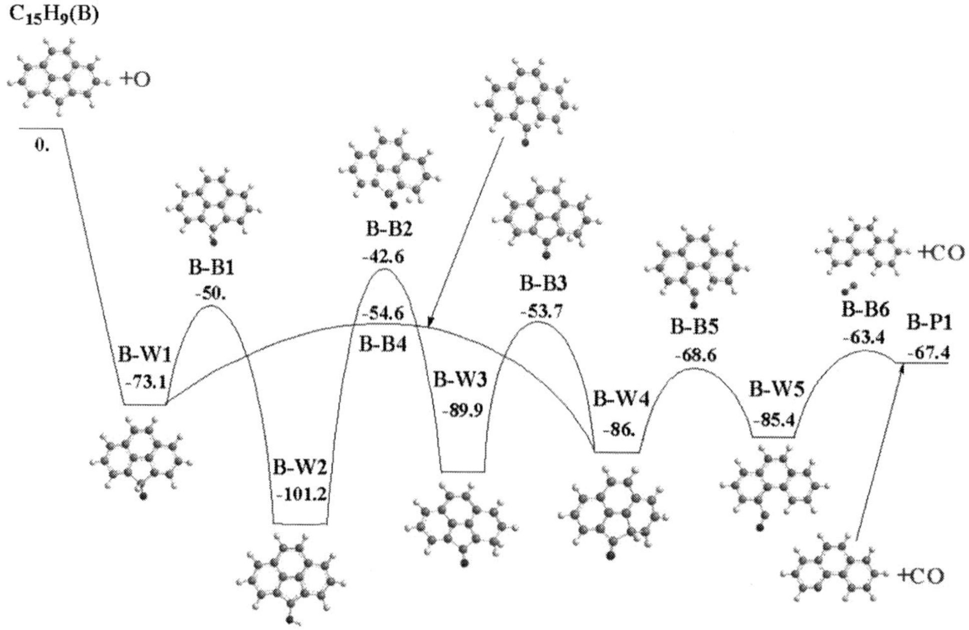

Fig. 2. Potential energy diagram for the $C_{15}H_9$(B) + O reaction calculated at the G3(MP2,CC)//B3LYP/6-311G(d,p) level of theory. All relative energies are given in kcal/mol with respect to the initial reactants.

In the $C_{15}H_9(B)$ + O reaction, the initial association step forming B_W1 is 73.1 kcal/mol exothermic. The decarbonylation process of B_W1 producing $C_{14}H_9$ + CO proceeds via the B_W2, B_W3, and B_W4 intermediates. B_W4 can be formed from B_W1 either in one step via B_B4, by 1,2-H migration to the *ortho* carbon with respect to the attacked C atom, or via a three-step pathway B_W1 → B_B1 → B_W2 → B_B2 → B_W3 → B_B3 → B_W4 involving consecutive 1,2-H shift from C to O, 1,4-H migration from O to the C atom in the neighboring six-member ring, and 1,2-H shift back to the *ortho* carbon atom. Here, the one-step process is clearly preferable since the transition state B_B4, 54.6 kcal/mol below the reactants, lies 12.0 kcal/mol lower in energy than B_B2. Opening of the five-member ring in B_W4 produces B_W2 via a relatively low barrier of 17.4 kcal/mol at B_B5 residing 68.6 kcal/mol below the reactants. Finally, B_W5 eliminates the CO group and produces the 4-phenanthrenyl radical $C_{14}H_9$ (B_P1). The critical transition state for the most favorable decarbonylation pathway is B_B4.

Conclusion

The most likely pathways of the reaction $C_{15}H_9$ + O were considered. The optimal geometry of the structures, vibrational frequencies, energies of products, reagents, intermediates and transition states that are involved in the $C_{15}H_9$ + O reaction were calculated.

The reaction pathways leading to the expulsion of the CO molecule were also determined. In the future, the calculated energetic and molecular parameters will be used to calculate pressure- and temperature-dependent rate constants for the oxidation of five-member rings with atomic oxygen on the surface of soot particles and large PAHs using the RRKM-Master Equation approach.

Acknowledgments

The work was supported by the Ministry of Education and Science of the Russian Federation under Grant no.14.Y26.31.0020 to Samara University.

References

1. H. Richter, J.B. Howard, Formation of polycyclic aromatic hydrocarbons and their growth to soot - a review of chemical reaction pathways, Prog. Energ. Combust. Sci. 26 565 (2000)
2. M. J. Frisch, G.W. Trucks, H.B. Schlegel, Gaussian 09, revision B.01. Wallingford, CT: Gaussian, Inc (2010)
3. H. J. Werner, P. J. Knowles, R. Lindh, MOLPRO, version 2015.1, a package of ab initio programs. Available at: http://www.molpro.net (2015)
4. Y. Georgievskii, J.A. Miller, M.P. Burke, and S. J. Klippenstein, Reformulation and solution of the master equation for multiple-well chemical reactions. J. Phys. Chem. A 117 12146 (2013)

Plasma vortex reactor for production of heat energy and hydrogen

Anatolii Klimov[1,2] *Svetlana* Kurushina[2], *Nonna* Molevich[2;3] , *Denis* Porfiriev[2;3] and *Igor* Zaversinskii[2*],

[1]Joint Institute for High Temperatures RAS, Moscow, 125412, Russia
[2]Samara National Research University, Samara, 443086, Russia
[3]Lebedev Physical Institute, Samara, 443011, Russia

Abstract. Numerical simulation of the turbulent three-dimensional swirling flow structure in plasma vortex reactor is conducted. Flow and set up parameters correspond to the experimental conditions. Flow velocity and thermal fields have been obtained. A qualitative agreement between the results of the calculations and the experimental data for pure argon has been shown.

1 Introduction

Recently, the problem of creating efficient compact energy sources, including alternative ones, has become increasingly noticeable [1,2]. One of the most promising technologies is the use of plasma vortex reactor (PVR), which can be both heat and hydrogen generator.

Intensive studies of a plasma chemical reactor started a few years ago [2-4]. Peculiar properties of gas-dynamics and heat fluxes in the presence of longitudinal electric discharge in the swirl flow were studied in [5,6]. The test gases used in this experimental set are the following: argon, argon + water vapor, argon + water vapor + nickel micro- and nanoparticles. The coefficient of performance was found. It was shown that COP depends on the parameters of the electric discharge and vortex flow itself. At the same time, the flow structure is determined by a number of factors: the shape and location of the discharge electrodes, swirl number, and the relative mass flow rates of the components.

A numerical simulation of the turbulent vortex flow structure in a model duct in the presence of a model heat source is carried out in [7]. The heat capacity of the swirler, electrodes and nozzle elements, as well as the thermal conductivity of the walls were neglected. It is shown that the recirculation zone forms in such system and its characteristics strongly depend on the heat source power and the shape of electrodes and their location. A qualitative agreement between the results of the calculations and the experimental data [8] has been obtained. The numerical modelling has demonstrated that it is possible to configure electrodes in such a way as to ensure energy transfer mainly downstream from the interelectrode zone. It may be accomplished by the combination of thin anode and pipe-like cathode through which areas of different pressures are connected.

* Corresponding author: ipzav63@mail.ru

Fig.1. Experimental setup. General view.

Here, a simulation gasdynamic and thermal fields structure formed in a 3D unsteady viscous turbulent vortex flow of argon in the experimental setup [8] is carried out.

2 Numerical modelling

2.1 Governing equations

Standard unsteady Reynolds averaged Navier-Stokes (URANS) equations were used to describe the flow:

$$\frac{\partial \rho}{\partial t}+\frac{\partial(\rho v_i)}{\partial x_i}=0,$$

$$\frac{\partial(\rho v_i)}{\partial t}+\frac{\partial(\rho v_i v_j)}{\partial x_i}=-\frac{\partial P}{\partial x_i}+\frac{\partial}{\partial x_i}\left[\mu\left(\frac{\partial v_j}{\partial x_i}+\frac{\partial v_i}{\partial x_j}-\frac{2}{3}\delta_{ij}\frac{\partial v_k}{\partial x_k}\right)\right]+\frac{\partial}{\partial x_i}\left[-\rho\overline{v'_i v'_j}\right] \quad (1)$$

$$\frac{\partial(\rho E)}{\partial t}+\frac{\partial[v_i(\rho E+P)]}{\partial x_i}=\frac{\partial}{\partial x_j}\left[\left(\kappa+\frac{c_p\mu_t}{\mathrm{Pr}_t}\right)\frac{\partial T}{\partial x_j}+v_i(\tau_{ij})_{eff}\right]+N(\vec{x}),\ P=\frac{\rho T}{M},$$

where $E=h-\frac{P}{\rho}+\frac{v^2}{2}$ is energy, $(\tau_{ij})_{eff}=\mu_{eff}\left(\frac{\partial v_j}{\partial x_i}+\frac{\partial v_i}{\partial x_j}\right)-\frac{2}{3}\mu_{eff}\frac{\partial v_k}{\partial x_k}\delta_{ij}$ is the deviatoric stress tensor; $\left[-\rho\overline{v'_i v'_j}\right]$ are the Reynolds stresses which must be modeled to close the set of equations, v_i, v_i', ρ, T P, E, and h are the mean and fluctuating velocity components, density, temperature, pressure, total energy, and enthalpy, respectively; N is the energy source, μ, μ_t, μ_{eff} are the molecular, turbulent, and effective viscosity coefficients, respectively; c_p is the molar specific heat capacity at constant pressure; κ is the thermal conductivity coefficient; and Pr_t is the turbulent Prandtl number.

The no-slip velocity and fixed temperature conditions were imposed along the tube and electrode surfaces. At the duct inlet and outlet, fixed mass flux conditions were used. Variation of the inlet turbulent intensity in the range of moderate values of 0-5% did not affect the flow structure discernibly. At the outlet, we imposed the boundary conditions with the static pressure equal to the atmosphere pressure.

Subsequently, we will solve the problem close to the experimental conditions [8], with ultimate goal of finding optimal configuration and location of electrodes.

2.2 Numerical simulation

The numerical simulation of the non-stationary 3D turbulent swirling flow was made using the ANSYS FLUENT 15.0 program package, which solves system (1) using the finite volumes method. In our simulation, the Spalart-Allmaras model was used. A detailed analysis of the use of the turbulence models in similar problems is given in [8].

For spatial discretization of density, momentum, energy and turbulent quantities, a second-order upwind scheme is applied. The use of higher-order schemes has not changed the flow parameters considerably. The diffusion terms are central-differenced and second-order accurate. We interpolate the pressure values at the faces using the PRESTO! scheme because of the strong swirl nature of the flow.

For transient terms, we used the fully implicit scheme of the second-order accuracy. Different pressure-velocity coupling schemes were tested and gave equal results. So, we used the SIMPLE scheme as the least resource intensive. The convergence was obtained when the residual reached 10^{-6} for the energy equation and 10^{-4} for the continuity equation, the momentum equation, and the equations for turbulent quantities.

The geometric modelling of the setup and the construction of the grid were carried out using the ANSYS Workbench package. The computational grid consisted of about $3 \cdot 10^6$ hexahedral cells, except for a small region, where the cells were tetrahedral, Fig 2.

Fig. 2. The computational grid near the swirler.

The grid refinement did not considerably alter the gasdynamic parameters. Calculations with different values of time step from $1 \cdot 10^{-4}$ sec to $1 \cdot 10^{-6}$ sec were carried out. Finally, the time step was set equal to $5 \cdot 10^{-5}$ sec since using lower values led only to an increase in the computational time.

The calculations have showed that the main features of the formation of the velocity and the thermal field are similar to founded in [7].

First, the structure of the flow is determined by the counterflow zone, typical for swirled flows [5,6], Fig. 3.

Secondly, using of two solid cylindrical electrodes leads to a significant drawback: between the electrodes, a dead zone is formed where the pressure and inertia forces almost prevent from the mass exchange with the outer flow. It leads to the interelectrode zone overheating and prevents for heat transfer downstream.

Thirdly, only using of thin anode and pipe-like cathode combination leads to optimal flow structure when the pressure gradient along the symmetry axis draws out the hot gas from the interelectrode area. The optimal electrodes location for this setup is close to the output of the PVR, Fig 4.

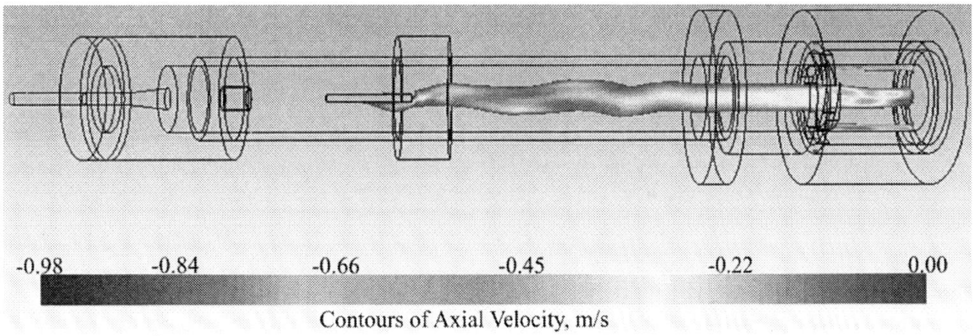

Fig. 3. Counterflow zone structure.

A hollow electrode is crucial for use in a PVP, since a solid one forms an obstacle that creates a negative axial pressure gradient and extends the counterflow zone. In this case, heat is effectively carried away downstream, so that the overheating zone isn't formed, the energy flux is directed mainly downstream from the cathode, Fig. 4.

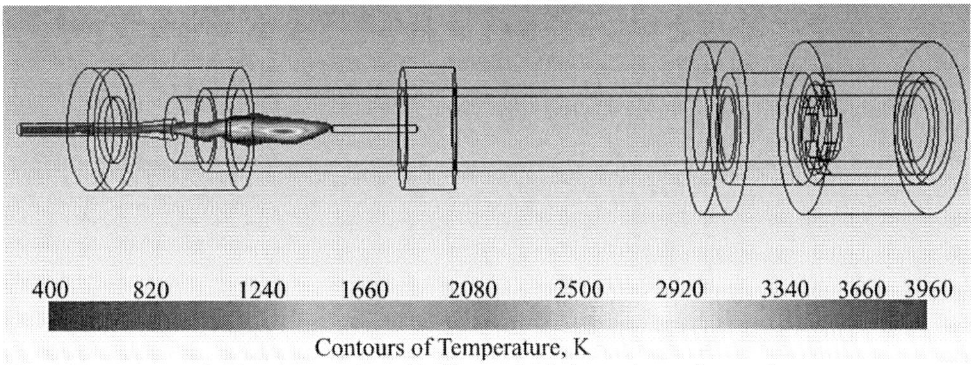

Fig. 4. Temperuture field at thin anode and hollow cathode located near the nozzle.

A qualitative agreement between the results of the calculations and the experimental data for pure argon [8] has been obtained. The next step in the simulation of the PVR is the calculation for real mixtures of Ni, H_2O, Ar, taking into account the plasma-chemical kinetics of the discharge mixture components as well as thermal characteristics of the experimental set up.

The study was supported in part by the Ministry of Education and Science of Russia under the public contract with educational and research institutions within the projects 3.1158.2017/4.6, 14.Y26.31.0020 and by RFBR under grant 16-41-630591.

References

1. M. Jasiński, H. Dors, J. Nowakowska, J. Mizeraczyk, Chem. Listy, **102** (2008).
2. A. Klimov, V. Bitiurin, B. Tolkunov, V. Chinnov, S. Godin, A. Efimov, D. Kutuzov and L. Polyakov, AIAA Paper 2011-3285 (2011).
3. V.A. Bitiurin, A.I. Klimov, O.V. Korshunov, V.F. Chinnov, High Temperature, **52** (2014).
4. V.A. Bitiurin, A.I. Klimov, O.V. Korshunov, V.F. Chinnov, High Temperature, **53** (2015).
5. A. Gorbunova, A. Klimov, N. Molevich, I. Moralev, D. Porfiriev, S. Sugak, I. Zavershinskii, International J. Heat and Fluid Flow, **59** (2016).
6. N. Syred, Prog. Energy. Combust. Sci., **32** (2006).

7. N.K. Belov, I.P. Zavershinskii, A.I. Klimov, N.E. Molevich, D.P. Porfiriev, B.N. Tolkunov. IOP Conf. Series: Journal of Physics: Conf. Series, **980** (2018).
8. P.N. Kazanskii, A.I. Klimov, N.E. Molevich, D.P. Porfiriev, I.P. Zavershinskii, IOP Conf. Series: Journal of Physics: Conf. Series **980** (2018).

CARS and Fluorescent study of ignition of H_2/O_2 mixtures upon photodissociation of O_2 molecular

Vitaly Kobtsev[1], *Sergey* Kostritsa[1*], *Dmitrii* Kozlov[2], *Alexey* Pelevkin[1,3], *Valery* Smirnov[2], *Natalia* Titova[1], *Sergey* Torokhov[1], *Konstantin* Vereshchagin[2], *Sergey* Volkov[2]

[1] Central Institute of Aviation Motors, Moscow, Russia

[2] Prokhorov General Physics Institute, Russian Academy of Sciences, Moscow, Russia

[3] Moskow Institute of Physics and Technology, Dolgoprudny, Moscow region, Russia

Abstract. The research is devoted to gas mixtures ignition by UV laser radiation. The dissociation of O_2 molecules by a pulse of excimer ArF-laser radiation at 193-nm wavelength with formation of the chemically active oxygen atoms initiating chain reactions which cause ignition of H_2/O_2 mixture was employed. The experimental test bench was created with CARS and fluorescent techniques for experimental investigation of some peculiarities of mixture ignition and combustion caused by such photo-dissociation, at conditions typical for combustion chamber. Two-dimensional numerical modeling of combustion process in model combustion chamber, based on kinetic mechanism of H_2 oxidation including atom $O(^1P)$ and radicals $OH(A^2\Sigma^+)$, was performed.

1 Introduction

Among different methods of gas mixture ignition, the considerable attention is paid to the burning initiation by laser radiation. In the overviews [1,2], the methods of resonant and non-resonant impact on the molecules of combustible gas mixture leading to heating, optical breakdown, photo-ionization or photo-dissociation are discussed. This work is devoted to the experimental and numerical study of temporal and space behavior of such important parameters of combustion development as temperature and concentration of radicals OH at the initial stage of ignition and combustion (~5–50 μs) of H_2-O_2 mixture upon the photo dissociation of O_2 molecules by the focused resonance laser radiation with a wavelength of 193 nm.

2 Experimental setup

For the investigation of photochemical ignition of gas mixtures at the pressure of 1–10 atm, the experimental setup was built. It consists of an excimer ArF-laser, a model combustion chamber provided with the glow plug for heating the gas mixture, and an optical diagnostic system. Fast processes in the chamber were studied using chemiluminescence emission (CLE) for detecting the electronically excited $OH(A^2\Sigma^+)$ radicals (this species is denoted

* Corresponding author: sergey.kostritsa@gmail.com

below as OH*), laser-induced fluorescence (LIF) for observation of OH radicals in the ground electronic state, and coherent anti-Stokes Raman spectroscopy (CARS) for measuring the temperature. The combustion of the preheated gas mixture was initiated by the pulse of the excimer ArF-laser, which irradiates at the wavelength of λ=193 nm. Such a radiation causes the excitation of O_2 molecules to the $B^3\Sigma_u^-$ electronic state with the following dissociation of these molecules on two oxygen atoms. The laser pulse duration is τ_p=15 ns, the energy in the pulse varies in the range E_s=25-150 mJ. Laser beam is focused by a lens with a focal length of 200 mm at the distance of 1 mm above the glow plug.

2.1 Temperature measurements

The gas mixture temperature was measured using both thermocouples and CARS techniques. At first, the temperature was measured by thermocouples at different pressure values in the case of absence of initiating laser radiation. For measurement of the temperature of the initial mixture and combustion products in the focus area of the excimer laser radiation both before and after the impact of excimer laser pulse, the broadband CARS spectrometer [3], which allows registering CARS spectra of a Q-branch of hydrogen molecules during one laser shot, was also used. The temperature can be determined in this way in the range of 500–3500 K with an accuracy of ±7%. The spatial resolution was ~1.7×0.15×0.15 mm^3. For the improvement of the CARS spatial resolution in the longitudinal direction, the scheme with the crossed beams radiated by the pump laser "planar CARS" [4], which were focused by the lens with the focus length of f=250 mm, was applied.

3 Modeling

3.1 O_2 molecule photo dissociation

For the analysis of the ignition and combustion process in the H_2-O_2 mixture under the exposure of laser radiation with the wavelength λ=193.3 nm, the numerical simulation was carried out. At the first stage, the absorption of laser radiation and the composition of the mixture in the irradiation region to the end of the laser pulse (τ_p=15 ns) were calculated. Computations shows that mostly oxygen atoms O(3P) form to the end of the pulse. Their mole fraction is maximal in the focal area due to the greatest values of temperature and radiation intensity in this region. According to calculations, the concentration of O atoms in the focal area can mount to 1% of total O2 concentration at maximal energy E_s ~150 mJ in the laser pulse.

3.2 Photochemical ignition of H_2-O_2 mixture

At the second stage, the numerical simulation of the ignition and combustion of H_2-O_2 mixture just after finishing the laser pulse was carried out with the usage of the program FLUENT from the ANSYS-CFD software package [5].

3.3 Kinetic mechanism

The reaction mechanism of H_2 oxidation [6] was used as basic one. This mechanism was supplemented by the set of reactions describing the formation of electronically excited radical OH* [7].

4 Results and discussion

4.1 LIF and CLE

The space-time distributions of OH and OH* radicals obtained during laser-induced ignition of the H_2/O_2 mixture with ϕ=1, T_0=713 K, P_0=1 atm and E_s=107 mJ have been studied. Two parameters of focusing of excimer laser radiation were used: a) in the chamber with the glow plug, the confocal parameter of focusing was equal to 8 mm, while b) in the chamber with the resistive heating, it was 24 mm. The results of experiments are compared in Fig. 1. In the first column (t=0 µs) are shown the images, which were registered at the moment of the UV laser radiation pulse and represent the luminescence of laser-excited $O_2(B^3\Sigma_u^-)$ molecules within the Schumann-Runge band in the focal area. The second column (t=10 µs) represents images corresponding to the induction period when the O_2 luminescence already finished, but the mixture burning did not start yet. The third images (t=30 µs) display the appearance of CLE signals of OH* radicals in the vicinity of the focal area, attesting that the ignition has occurred. Further evolution of the chemiluminescence zone is illustrated by the images with t=50 and 80 µs. It is seen that in the case b), ignition starts simultaneously with a) but in a larger area. It is evident that a total time needed for complete combustion of the mixture will be smaller in the case b) than that in the case a). It means that using the excimer laser radiation, one can ignite simultaneously a large volume of a flammable mixture.

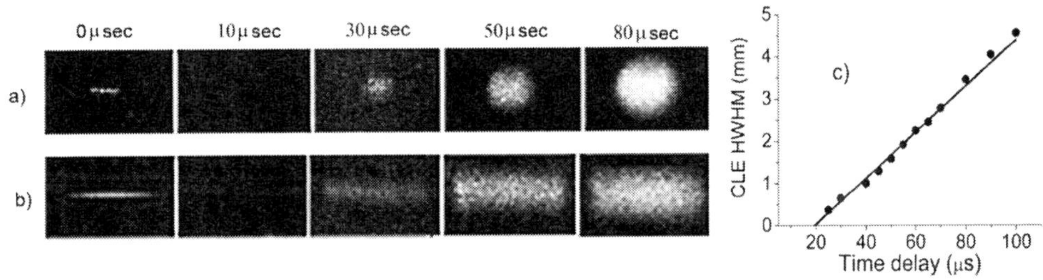

Fig. 1. a), b) The sequences of images of a chemiluminescence area of the OH* radicals depending on the time delay from the excimer laser pulse in a stoichiometric H_2/O_2 mixture with P_0=1 atm and E_s=107 mJ. c) Sizes of chemiluminescence area at different delay time.

The obtained images a) were processed as follows. The half-width at half-maximum $(I_{ch})_{max}/2$ (HWHM) of the chemiluminescence intensity profile in the observed region was determined at each time delay and presented at graph (c picture). The point of intersection of the straight line $r_{ch}(t)$ with the abscissa axis (in this point the combustion region seems to converge to zero size) can be treated as the ignition delay (induction) time t_{ind} (see Fig. 1 c).

4.2 CARS thermometry

The evolution of temperature in the focal region was measured using coherent anti-Stokes Raman spectroscopy (CARS).The results of temperature measurements in the focal region as a function of time for different values of ϕ are given in Fig. 2.

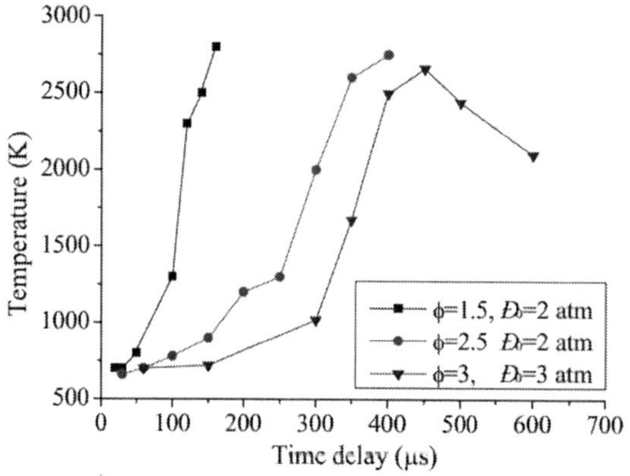

Fig. 2. Evolution of gas temperature in the focal zone profiles during the ignition of the H_2-O_2 mixture with different values of ϕ, T_0=713 K and E_s=107 mJ.

These dependencies show that, after a certain delay (induction period), the temperature rapidly increases up to value of ~2700 K. The decrease of the temperature at large values of t (this is shown for the case with ϕ=3) is due to the cooling of the gas due to the heat release through the chamber walls. One can also see from Fig. 2 that the induction time t_{ind} increases with increasing value of ϕ. In Fig. 3a is shown the comparison of the measured and calculated values of the temperature in the focal zone during the ignition of the H_2-O_2 mixture with ϕ=1.3, T_0=713 K, P_0=2 atm and E_s=105 mJ. One can see that the calculation results describe the dynamics of the temperature change in this region sufficiently well. A weak linear growth in calculated temperature is observed at the initial stage ($t \leq 35$ μs). At this stage, O atoms, produced by laser radiation, initiate the chain reactions, which lead to a gradual increase in the concentration of O, H atoms and OH radicals and to a slow release of heat due to the H_2O formation (see Fig. 3b). After the concentrations of active components reach a certain value, the mixture ignites. This is characterized by a sharp increase in the H_2O concentration and temperature, which reaches the value of ~2800 K to the time t=60 μs both in the calculation and in experiment.

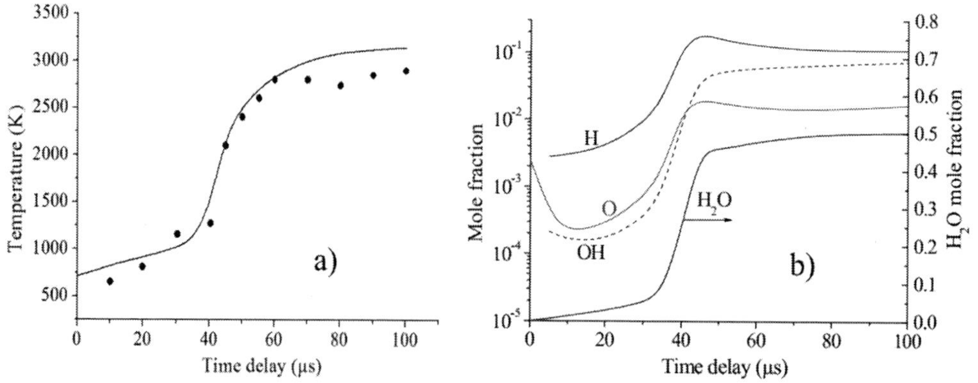

Fig. 3 Measured (points) and computed (curve) evolution of gas temperature (a) species mole fractions (b) at the focal zone during the ignition of the H_2-O_2 mixture with ϕ=1.3, T_0=713 K, P_0=2 atm and E_s=105 mJ.

The difference between the calculated and measured temperature at $t \leq 35$ μs is likely related to the above-described features of temperature measurements when the size of the high-temperature zone is less than the spatial resolution of CARS. At t~60 μs, the mixture burns out in the focus area and the gas temperature changes only slightly. The calculated temperature of the combustion products turned out to be a little higher than measured one due to the neglect of the heat losses through the chamber walls during modeling.

Conclusions

An experimental setup was created allows to investigate the ignition process of various gaseous fuels with oxygen, the reactivity of which is instantly (within 20 nsec) activated in relation to the chain reactions by oxygen atoms obtained as a result of dissociation of O2 molecules. This type of ignition does not contain impurities of the recombining plasma characteristic of spark methods of optical or electrical breakdown.

Based on the space-time distributions of OH and OH* radicals obtained during laser-induced ignition of H_2-O_2 mixture at the mixture temperatures below the autoignition temperature, the quantitative data of the induction time and the velocity of the flame front were obtained for various parameters of the mixture and different values of the energy in the laser pulse.

The temperature measurements by CARS-spectroscopy has allowed to getting information about the temperature evolution at the initial stage of the burning process register experimentally the presence of high temperature $\geq$ 1000 K at time delays starting from 30 μsec and to confirm the fact of mixture ignition with transition to steady burning, obtained on the basis of observation of the emission of OH and OH* radicals.

References

1. P.D. Ronney, Laser versus conventional ignition of flames, Optical Engineering **33**(2), 510–521 (1994)
2. S.A. O'Briant, S.B. Gupta, S.S. Vasu, Review: laser ignition for aerospace propulsion, Propulsion and Power Research **5**(1), 1–21(2016)
3. V.V. Smirnov, S.A. Kostritsa, V.D. Kobtsev, N.S. Titova, A.M. Starik, Experimental study of combustion of composite fuel comprising *n*-decane and aluminum nanoparticles, Combust. Flame 162(10) 3554–3561 (2015)
4. A.C. Eckbreth, BOXCARS: Crossed-beam phase-matched CARS generation in gases, Appl. Phys. Letts. **32** ,421-423 (1978)
5. ANSYS FLUENT. Version 12. ANSYS Inc.; January (2009)
6. Bezgin L.V., Kopchenov V.I., Sharipov A.S., Titova N.S., Starik A.M. Evaluation of prediction ability of detailed reaction mechanisms in the combustion performance in hydrogen-air supersonic flows. Combustion Science and Technology, 2013, V.**185**, N.1, P. 62-94 (2013)
7. T. Kathrotia, M. Fikri, M. Bozkurt, M. Hartmann, U. Riedel, C. Schulz. Study of the H+O+M reaction forming OH*: Kinetics of OH* chemiluminescence in hydrogen combustion systems. Combustion and Flame 157 (2010) 1261–1273(2010)

Investigation of thermal degradation of hydrolysis lignin

Pavel Maryandyshev[1,*], *Aleksei* Kangash[1], and *Victor* Lyubov[1]

[1]Northern (Arctic) Federal University, Naberezhnaya Severnoy Dviny 17, Arkhangelsk, 163002, Russia

Abstract. Wood hydrolysis facilities were built in both northern and southern European regions of Russia, as well as in Siberia and the Russian Far East. Other hydrolysis plant in the former USSR were located in Ukraine, Belarus, Uzbekistan, Kazakhstan, Moldova and Lithuania. All or most of these facilities produced substantial amounts of excess lignin, so-called "hydrolysis lignin" (HL). Technology of energy utilization of HL, having a great potential is a production of granulated fuel. Perspective way is pelletization of HL. In this paper complex kinetic analysis of hydrolysis lignin and its organic components under conditions of dynamic and isothermal heating was carried out. Kinetic parameters are determined using the extended independent parallel model.

1 Introduction

Recently more attention is being paid to wood biofuels, which are renewable energy sources and take the fourth position in the world energy balance after coal, natural gas and oil that are connected with some factors. One of these factors is development of wood biomass energy utilization technologies. Thus, there is a huge possibility of utilization of side product of wood treatment – lignin which is 40% of dry mass of wood [1].

In the technological cycle of hydrolysis industry lignin is being produced, which is a complex of components of different chemical structure. Hydrolysis lignin is composed of: mainly of changed lignin of vegetative cellulose (40–88% of dry mass); polysaccharides (13–45%); components of lingo humic complex (5–19%); non-washed monosaccharides after hydrolysis process; mineral and organic acids: mainly sulphur, formic and acetic, ashes and some other components [2, 3].

Acid hydrolysis of wood and agricultural residues for the production of ethanol, fodder yeast, hexose and pentose sugars and their derivatives (i.e. furfural and xylitol) had industrial applications in Soviet Union and Bulgaria [4]. It was based mainly on percolation hydrolysis using diluted H_2SO_4, which was developed in the former USSR and used from the 1930s to the 1990s, although some semi-industrial facilities for wood or agricultural residue hydrolysis using concentrated HCl were also used during this period. Wood hydrolysis facilities were built in both northern and southern European regions of Russia, as

* Corresponding author: p.marjyandishev@narfu.ru

well as in Siberia and the Russian Far East. Other hydrolysis plant in the former USSR were located in Ukraine, Belarus, Uzbekistan, Kazakhstan, Moldova and Lithuania. All or most of these facilities produced substantial amounts of excess lignin, so-called "hydrolysis lignin" (HL). In spite of the fact that there were energy utilizing boilers, they were not able to use all hydrolysis lignin for the energy production and main part of HL was stored in dumps.

Technology of energy utilization of HL, having a great potential is a production of granulated fuel [1]. Perspective way is pelletization of HL. And combination of the preliminary torrefaction process of HL before its pelletization (isothermal heating at 250–300°C without oxygen) of the initial HL will lead to receiving a new product with better heating values: energy yield and density. First experience of such plant in the Arkhangelsk region of the Russian Federations shows that pellets from HL has calorific value not lower than 21,34 MJ/kg (with W_r=6,27%; A_r=2,45%).

Therefore, a huge interest is being raised to the energy utilization of the hydrolysis lignin in the Russian Federation. For the effective energy utilization of the HL it is necessary to study deeply its characteristics, process of thermal degradation and combustion, to analyze gas components and ashes during the process of thermal degradation, to carry out comparison with classical wood fuel (spruce, pine). In this paper thermogravimetric curves were constructed, gas components were determined, and kinetic parameters were calculated using an extended independent parallel model. The results were compared with parameters of classic wood fuels.

2 Materials and methods

Wood fuel of various species (spruce and pine trunk, spruce and pine bark, hydrolysis lignin) were used during the investigation for comparisons.

The content of carbon, hydrogen, nitrogen and sulphur was determined using the EuroVectorEA-3000 analyser, and oxygen was calculated from the mass balance.

The thermotechnical characteristics of the test samples were determined in accordance with State Standard 54185-2010, 54186-2010, and 54211-2010. Net calorific values were determined using an IKA C 2000 Basic version 2 bomb calorimeter.

For a deeper understanding of the processes of thermal decomposition and combustion of wood fuels, it is necessary to know what components they consist of. For this, X-ray fluorescence analysis was performed.

Samples were dynamically heated to produce thermogravimetric curves. Experiments were carried out using a synchronous thermal analyzer Netzsch STA 449 F3.

For isothermal heating 250-mg sample was placed in the crucible of the reactor with a special spoon and heated from ambient temperature (20°C) to a temperature of 550°C; then, it was kept under isothermal conditions for an hour under air, which was obtained mixing of nitrogen with oxygen (Fig. 1). The analyser of gas components (Anapol EU 5000), which consists of electrochemical cells for the determination of CO, CO_2, O_2, and total hydrocarbons, allowed to perform continuous measurements.

Kinetic parameters were calculated using extended independent parallel model (EIP). The EIP model combines the parallel processes of thermal degradation of the three organic components of wood biomass (hemicellulose, cellulose and lignin). To implement this method, the ratio coefficients for hemicellulose, cellulose and lignin must be determined, which is not an easy task [5]. In this paper, the desired relations were determined from differentiated thermogravimetric curves and were calculated from the results of modelling.

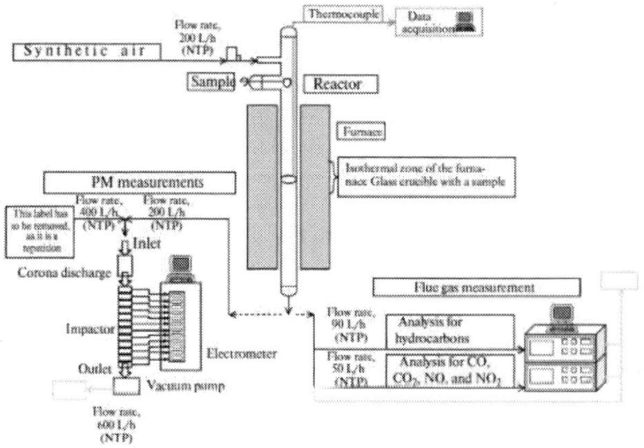

Fig. 1. Schematic diagram of the experimental installation for the determination of gas components and fine particles emitted during the thermal decomposition and combustion.

3 Results and discussion

The elemental composition of hydrolytic lignin is shown in Table 1.

Table 1. The elemental composition of hydrolytic lignin.

	C (%)	H (%)	N (%)	S (%)	O (%)
Hydrolysis lignin	52,6	5,8	0,2	-	29,8

The results of proximate analysis are shown in Table 2.

Table 2. Proximate analysis.

Sample	Moisture W_a (%)	Ash content A_a (%)	Content of volatile organic component V_a (%)	Net calorific value Q_i^r (kJ/kg)
Hydrolysis lignin	8,6	3,0	57,8	19315
Spruce (trunk)	6,9	1,0	79,1	17028
Spruce (bark)	11,6	2,2	65,5	16982
Pine (trunk)	7,4	0,3	79,1	17313
Pine (bark)	9,5	2,5	68,6	17292

Classical TG and DTG curves in Figures 2–3 describe the evaporation of moisture and volatile substances. Considering hydrolysis lignin, it can be distinguished 3 main zones. The first zone, in the temperature range from 20 to 140 °C with a weight loss of about 10% is characterized by evaporation of moisture. The second zone is the zone of active pyrolysis, corresponds to a temperature range of 250–400 °C. In this zone, the main mass loss occurs (from 45 to 60% depending on the type of fuel), the thermal degradation of hemicellulose and cellulose that make up the components of wood biofuels occurs. The third zone is in the temperature range from 400 to 800 °C and characterizes the process of passive pyrolysis when there is a prolonged thermal degradation of organic substances, mainly lignin. The mass loss in this range is from 10 to 15%, depending on the sample.

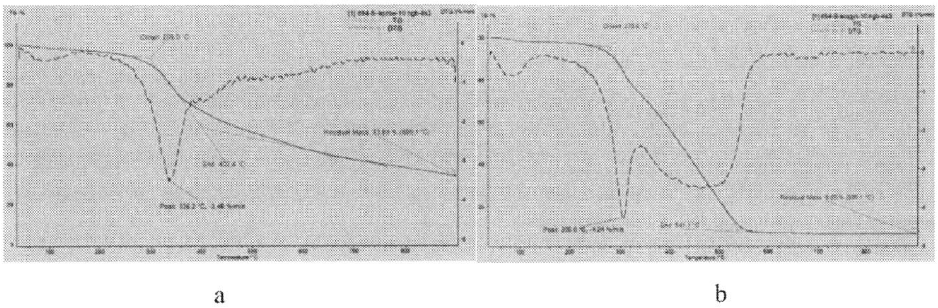

Fig. 2. TG and DTG curves for the thermal degradation of hydrolysis lignin under inert atmosphere (a) and under air (b)

Figure 3 illustrate data on the gas components CO, CO2, and hydrocarbons for hydrolysis lignin: the decomposition of the organic components of the test materials begins at a temperature of about 200°C. The release of gases reaches a maximum at 380– 400°C with the simultaneous combustion of volatile substances and of the coke residue. A lower volatile content and a considerably higher ash content in the hydrolytic lignin are responsible for a lower concentration level of the resulting gaseous components (to 8000 ppm) than that in the case of the spruce wood (to 11000 ppm). Note that carbon dioxide predominated in the resulting gaseous components.

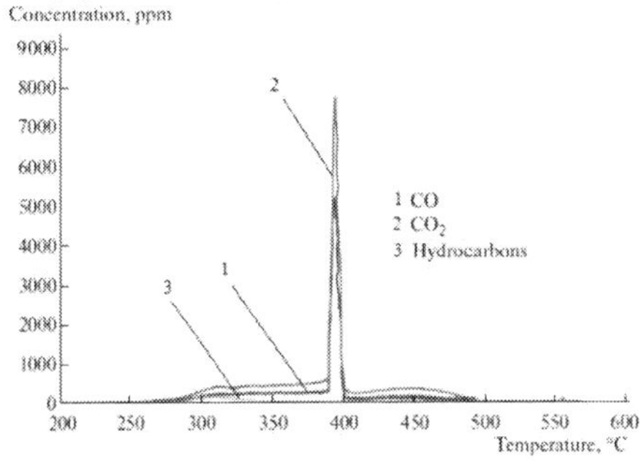

Fig. 3. Gas components formed in the course of the thermal decomposition and combustion of hydrolysis lignin

Table 4 shows the kinetic parameters for hydrolysis lignin in the inert and air atmosphere, determined by the EIPR model. This model is being mainly developed by the scientists of the Laboratory of Risk Analysis of the University of Haut Alsace [7].

Table 4. Kinetic parameters for hydrolysis lignin.

Sample	Kinetic parameters			
	Nitrogen atmosphere		Air atmosphere	
	A (s^{-1})	E_α (kJ/mol)	A (s^{-1})	E_α (kJ/mol)
Hemicellulose	$1{,}8\cdot10^4$	78,0	$1{,}8\cdot10^4$	79,0
Cellulose	$1{,}1\cdot10^4$	87,0	$4{,}1\cdot10^7$	115,0
Lignin	10,1	60,0	10,1	50,0
Coke residue	-	-	$6{,}0\cdot10^4$	99,0
Error	0,013		0,035	

4 Conclusions

Complex kinetic analysis of hydrolysis lignin and its organic components under conditions of dynamic and isothermal heating was carried out. Hydrolysis lignin has good composition elements and results of the proximate analysis are quite similar to results of the classical wood. Kinetic parameters are determined using the extended independent parallel model. The obtained experimental data allowed to conduct comparison of calculated results with experimental ones using thermogravimetric and differential-thermogravimetric curves. The error obtained with the use of the extended independent parallel model has lower values than for the differentially isoconversional method. Next step is the optimization of the process of thermal degradation and combustion in heat generating equipment using kinetic parameters.

References

1. Rabinovich, M.L., Fedoryak, O., Dobele, G., Andersone, A., Gawdzik, B., Lindstrom, M.E., and Sevastyanova, O., Renew. Sust. En. Rev., **57**, 1009 (2016)
2. Lyubov, V.K. and Lyubova, S.V., *Povyshenie effektivnosti energeticheskogo ispol'zovaniya biotopliv (Enhancement of the Efficiency of the Power-Generation Use of Biofuel)* (Arkhangelsk, 2010)
3. Mahmood, N., Yuan, Z., Schmidt, J., and Xu, C.C., Biores. Technol., **74**, 4 (2015)
4. Rabinovich, M., L, Cell. Chem. Technol., **44**, 174 (2010)
5. Rueda-Ordonez, Y.J. Tannous. *Isoconversional kinetic study of the thermal decomposition of sugarcane straw for thermal conversion processes* (Bioresource technology, 2015)
6. Carrot, P.J.M., Suhas, A., Carrot, M.M.L., Guerrero, C.I.,and Delgado, L.A., J. Anal. Appl. Pyrol., **82**, 264 (2008)
7. Popova, E., Chernov, A., Maryandyshev, P., Brillard, A.,Kehrli, D., Trouve, G., Lyubov, V., and Brilhac, J.-F., Biores. Technol., **218**, 1047 (2016).

Experimental study of chemiluminescence in UV and VIS range at hydrogen-oxygen mixtures ignition

Alexander Eremin[1] , *Mayya* Korshunova[1,2] , and *Ekaterina* Mikheyeva[1,2,*],

[1]JIHT RAS, 125412 Moscow, Russia
[2]BMSTU, 105005 Moscow, Russia

Abstract. The nonequilibrium radiation in the spectral range of 210-415 nm at ignition of a 10% stoichiometric hydrogen-oxygen mixture with additives of combustion inhibitors diluted with argon behind shock waves was registered. The detected chemiluminescence is presumably attributed to electronically excited H_2O^* and $H_2O_2^*$. Instead of the expected quenching of excited radicals and molecules in the ignition zone, with the addition of halogenated hydrocarbons inhibitors, the increase of radiation, particularly in the range of 330-415 nm, was observed. The possible reasons of this phenomenon are discussed.

1 Introduction

It is well known that the zone of ignition of hydrogen-oxygen mixtures is accompanied by the intense UV emission, usually attributed to excited OH* radicals. At the same time the other electronically excited radicals and molecules (HO_2*, H_2O*, H_2O_2*) can make an important contribution to the observed emission spectra [1-4]. It is assumed that the intensity of overequilibrium emission at ignition is an indicator of the rate of development of the ignition process. Therefore, when studying extremely topical problems of inhibiting the ignition of hydrogen, recording of nonequilibrium UV emission is usually one of the main diagnostic methods. In previous works [5, 6], it was shown that halogenated hydrocarbons are the effective inhibitors of hydrogen ignition and substantially increase the ignition delay times of hydrogen-oxygen mixtures behind shock waves. In a recent paper [7] it was suggested that the one of the possible inhibition mechanisms is the resonance quenching of excited states of active radicals, in particular, the peroxyl radical HO_2*, which plays a key role in chain branching upon ignition of hydrogen-oxygen mixtures $H + O_2 \rightarrow HO_2{}^* \rightarrow O + OH$, $H + O_2 + M \rightarrow HO_2 + M$ [8, 9], in collisions with molecules of inhibitors.

Therefore, the goal of this work is to analyse the effect of halogen-containing combustion inhibitors (CCl_4, $C_2F_4Br_2$) addition, which effect on hydrogen ignition was previously studied in detail by the authors [5, 6], as well as inert CO_2 addition, on the level of nonequilibrium UV radiation in spectral range from 210 up to 415 nm at ignition of

* Corresponding author: ekaterina.mikheyeva@gmail.com

stoichiometric hydrogen-oxygen mixture with additives of combustion inhibitors behind shock waves.

2 Experimental

The experiments were carried out in a standard shock tube equipped with modern optical diagnostic methods. The ignition behind the reflected shock wave was detected with the use of piezoelectric pressure sensors, as well as nonequilibrium radiation was detected with a photomultiplier, at 310±5 nm (which corresponds to the electronic transition of the OH * radical). Spectral emission at wavelengths from 210 up to 415 nm was measured with the help of an Acton Research Cooperation SpectraPro 150 diffraction monochromator and photomultiplier Hamamatsu H6780-04 with spectral resolution of ±4 nm. The spectral resolution is checked using mercury line measurements. The quantum efficiency of the entire setup is calibrated using a tungsten lamp with a known spectral emissivity. The investigated spectral range is characteristic for the emission of electronically excited radicals and molecules HO_2*, H_2O*, H_2O_2* [10, 11]. The parameters of the mixtures studied, as well as the ranges of temperatures and pressures behind the reflected shock waves, are given in the table 1.

Table 1. The experimental mixtures and parameters.

Mixture title	H_2, %	O_2, %	CO_2, %	CCl_4, %	$C_2F_4Br_2$, %	T, K	p, bar
Mixture1	6.7	3.3	--	--	--	1030-1400	4.3-5.8
Mixture2	6.7	3.3	10	--	--	1045-1300	4.3-5.5
Mixture3	6.7	3.3	--	2	--	1180-1400	4.3-4.4
Mixture4	6.7	3.3	--	--	1	1130-1400	5.6-5.8

Examples of characteristic oscillograms are presented in Fig. 1 - pressure profiles (a) and nonequilibrium radiation in the 310 nm (b) and 411 nm (c) regions (for another wavelengths the signal is similar), upon ignition of H_2/O_2 mixtures without an inhibitor and with the addition of 2% CCl_4 at similar temperatures behind shock waves. At the moment of ignition, characterized by a sharp rise in pressure, a peak emission signal appears at all wavelengths.

3 Results

It is clearly seen from the Fig.1 the addition of an inhibitor does substantially increase the ignition delay and, at the same time, reduce the intensity of nonequilibrium emission of the OH* radical at 310 nm. However, in the region of 411 nm, on the contrary, there is a significant increase in nonequilibrium emission.

Similar oscillograms were obtained in all the mixtures studied at different temperatures from 1050 to 1400 K. In additional series of experiments the absorption of intense radiation from a gas discharge lamp DKSSh-150 (having a brightness temperature of about 4000 K in these spectral regions) against a background of the signal was measured. Additional emission was detected from the source, which indicated a highly super-equilibrium character of the observed emission, with an effective excitation temperature above the brightness temperature of the source.

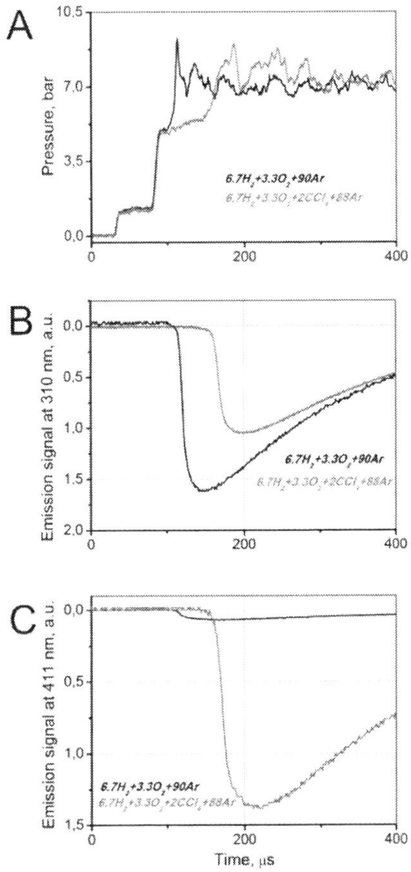

Fig. 1. Characteristic oscillograms of pressure profiles (a), and nonequilibrium radiation at 310 nm (b) and 411 nm (c) regions, at ignition of H_2/O_2 mixtures without an inhibitor ($p_5 = 5.2$ atm, $T_5 = 1215$ K) and with 2% CCl_4 ($p_5 = 5$ atm, $T_5 = 1250$ K).

The spectral dependencies of detected nonequlibrium radiation are shown in Fig. 2. It is seen from this figure that at 310 nm, corresponding to the chemiluminescence signal of the OH * radical, in the temperature range studied, the presence of additives leads to a slight decrease in the observed amplitude of the signal in 2 - 3 times, while at a wavelength of 210 nm, with the addition of CO_2, the amplitude of the signal did not change, and with the addition of $C_2F_4Br_2$ and CCl_4, an increase in the signal of emission was observed almost the tenfold. But the most significant increase in nonequilibrium radiation was observed in a spectral range of 330-415 nm, where all additives, including inert CO_2, led to an almost 20-fold increase in the radiation intensity.

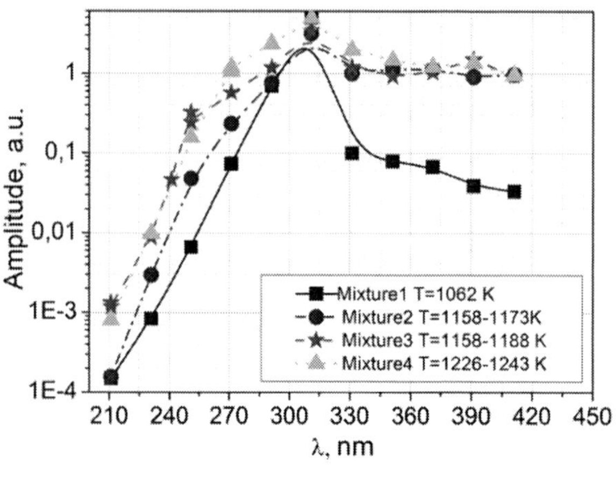

Fig. 2. Nonequlibrium emission spectra detected at ignition of hydrogen-oxygen mixtures.

4 Discussion

To analyse the possible mechanisms of amplification of nonequilibrium UV radiation with the addition of inhibitors to the hydrogen-oxygen mixture, numerical simulation of the kinetics of ignition of the investigated mixtures was carried out using the Chemkin code and kinetic mechanisms proposed in [12] for a hydrogen-oxygen mixture without additives and with the addition of CO_2, and in a recent paper [5] - for mixtures with CCl_4 additions. Both mechanisms are written in the form of "formal" kinetics, suggesting equilibrium in internal degrees of freedom of molecules. Note that for $C_2F_4Br_2$ no detailed kinetic combustion mechanism has been developed, therefore no numerical simulation has been performed.

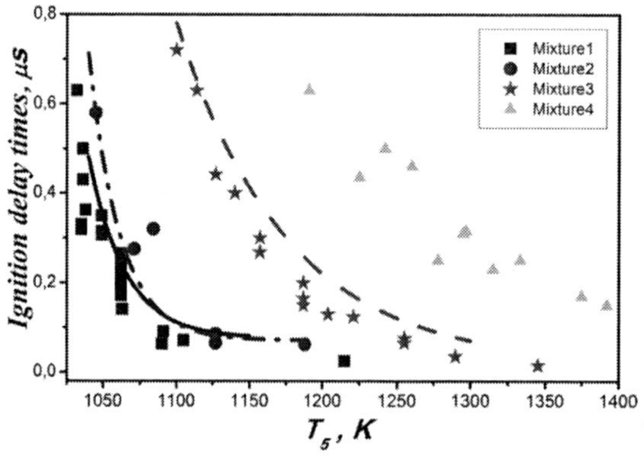

Fig. 3. Temperature dependences of ignition delay times. Points – experimental data, curves – the results of kinetic simulation.

The results of the calculations have shown that the theoretical model describes well the experimentally measured increase in ignition delay times with the addition of inhibitors (see Fig. 3). The addition of the CCl_4 inhibitor leads to a significant decrease in the maximum concentration of the HO_2 radical (in the ground state) and slightly increases the maximum concentration of hydrogen peroxide H_2O_2. HO_2 is first consumed with addition of inhibitor in the reactions $Cl+HO_2 \rightarrow HCl+O_2$; $CCl_3+HO_2 \rightarrow COCCl_2+Cl+OH$ and secondly, it is less formed in the reaction $H+O_2(+M) \rightarrow HO_2(+M)$ due to consumption of O_2 through the reactions $Cl+O_2=ClO+O$; $CCl_3+O_2=COCl_2+ClO$. Thus, the observed chemiluminescence could hardly be associated with overexcited hydroperoxyl radicals. Hydrogen peroxide and water with addition of inhibitor are formed mainly through reactions $OH+OH \rightarrow H_2O_2$; $OH+OH \rightarrow H_2O+H$. The production of OH is in turn connected with atomic hydrogen via $H+HO_2 \rightarrow OH+OH$; $H+O_2 \rightarrow O+OH$, whose formation is promoted by the reaction $Cl+H_2 \rightarrow HCl+H$.

Thus, the observed increase in the intensity of chemiluminescence can be conventionally attributed to an increase in the concentration of electronically excited molecules $H_2O_2^*$ or H_2O^*, which may indicate a redistribution of the reaction channels, but of course, in any way cannot lead to the observed effect of inhibition.

5 Conclusion

In this study, for the first time, a significant increase in the intensity of nonequilibrium radiation in the range of spectra 210-415 nm at ignition of hydrogen in the presence of inhibiting additions of halogenated hydrocarbons, was experimentally observed. Thus, the assumption of a mechanism for inhibiting the ignition of hydrogen by quenching excited radicals [7] is hardly justified, and the real mechanism of inhibition, as shown in [5, 6], is the capture of the primary radicals by inhibitors.

This work was supported by the grant of the RNF - № 14-19-00025.

References

1. T. Fiala, T. Sattelmayer, J. Propul. Power **33**, 490 (2016)
2. R.W. Schefer, W.D. Kulatilaka, B.D. Patterson, Combust. Flame **156**, 1234 (2009)
3. M. Vanpee, R.J. Mainiero, Combust. Flame **34**, 219 (1979)
4. P.J. Padley, Trans. Faraday Soc. **56**, 449 (1960)
5. A. Drakon, A. Eremin, N. Matveeva, E. Mikheyeva, Combust. Flame **176**, 592 (2017)
6. A. Drakon, A. Eremin, Combust. Sci. Technol. **190**, 550 (2018)
7. O. Skrebkov, Combust. Theor. Model. **19**, 138 (2015)
8. R.W. Getsinger, G.L. Schott, J. Chem. Phys. **43**, 3237 (1965)
9. B.B. Darwent, V.J. Krasnansky, Symp. (Int.) Combust. **7**, 3 (1958)
10. Z. Hong, K.-Y. Lam, R. Sur, S. Wang, D.F. Davidson, R.K. Hanson, Proc. Combust. Inst. **34**, 565 (2013)
11. H. Kijewski, J. Troe, Int. J. Chem. Kin. **3**, 223 (1971)
12. Z. Hong, D. Davidson, R. Hanson, Combust. Flame **158**, 633 (2011)

Environmental advantages of composite fuels based on industrial wastes and different ranks of coal

Galina Nyashina[1,*]

[1]National Research Tomsk Polytechnic University, Tomsk, 634050, Russia

Abstract. Thermal power plants and boiler units generate most of the anthropogenic emissions around the world. A promising solution to many problems that heat and power industry is facing today would be switching from conventional coal dust combustion to composite liquid fuels (CLF). These are also known as coal-water slurries containing petrochemicals (CWSP). Here, we perform an experimental study of the most hazardous anthropogenic emissions (sulfur and nitrogen oxides) from the combustion of high-potential CWSP. We identify the main benefits and potential drawbacks of using CWSP in heat and power industry. A set of components and additives to CWSP are explored that significantly affect the environmental and energy performance of fuels. The anthropogenic emissions from the combustion of CWSP made of widespread coal and oil processing wastes are no higher than those from coal dust combustion. Using specialized additives to CWSP, we can change the concentrations of NOx and SOx several times. The most appealing additives to CWSP are sawdust, straw, charcoal, limestone, and glycerol.

1 Introduction

Fossil fuels are the primary sources of heat and electricity in many countries. In the global energy mix, they account for over 80% [1–3], 20% of which is covered by coal. World energy production relies heavily on coal-fired thermal power stations and boiler plants. New coal-burning power stations and boiler plants appear every year, despite severe environmental implications. As the energy demand increases, coal-fired thermal power stations and boiler plants will only reinforce their leading positions in the coming decades. For instance, a report by the International Energy Agency [4] states that almost 41% of the world's electricity in 2013 was generated using coal. This number went up to 46% [4] in 2014–2016 and is expected to reach 50–52% in 2017–2020.

The mass usage of coal causes severe adverse environmental impacts and disrupts the ecological balance [5, 6]. Coal combustion leads to high emissions of nitrogen- and sulfur-containing oxides (NO_x and SO_x), volatile organic compounds (ash, soot, and dust), carbon dioxide (CO_2), as well as various traces of metals that can dissipate in the atmosphere over vast areas and have a significant negative effect on human health [7, 8].

* Corresponding author: gsn1@tpu.ru

The one considered the most promising is boosting the coal combustion efficiency by producing coal-based composite fuels: coal-water slurries with or without petrochemicals. Such an alternative to the conventional use of coal does not only alleviate global environmental problems but also provides a significant effect in terms of economy and energy performance [9–11]. In particular, involving low-rank coals and numerous coal and oil processing wastes into the process expands the scope of raw materials for the heat and power industry [11].

Coal-water slurry fuel (a mixture of water, coal dust, and chemical additives) is a viscous, stable, fire- and explosion-safe slurry, which is environmentally friendly at all stages of production and usage. Its fuel base consists of power-generating coals as well as non-power coals and coal sludge with a concentration of solid substances of no less than 60% [12]. Production of composite power fuels is the next step in the development of coal-water slurry technologies.

At the same time, there are some problems with the implementation of CWS technologies. Coal-water slurries are inferior to high-carbon fuels in the proportion of carbon. For this reason, they are notable for lower combustion temperatures and heat release (heat of combustion). Moreover, their erratic ductility makes it necessary to use chemical additives and stifles their advancement. This problem can be resolved by adding even as little as 10–15 wt.% of liquid fuel component to coal-water slurry fuels. This is how coal-water slurries containing petrochemicals are produced [11, 12]. The following substances can serve as liquid fuel components: waste industrial and motor oils, flammable liquids, emulsions, slurries, industrial wastes, etc. The range of these raw materials is growing rapidly [11].

The objective of this work is to undertake a comprehensive study of the environmental consequences brought about by the combustion of CLF slurries and coals of different ranks with high-potential additives.

2 Experimental setup and methods

To study the anthropogenic emissions from the combustion of coals as well as high-potential CWS and CWSP, we used an experimental setup similar to the one used in [13]. The key elements that ensured the continuous experimental studies included a combustion chamber, gas analyzing equipment, and high-speed recording devices (fig. 1).

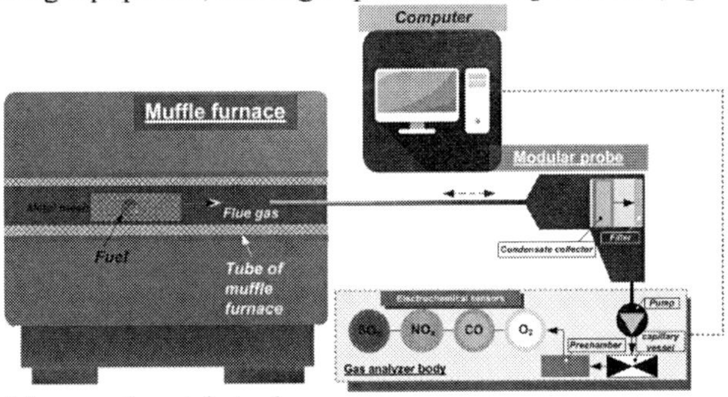

Fig. 1. The scheme of the experimental stand.

The method we used to determine the amount of emissions from fuel combustion involved the following procedures: preparing a fuel sample, placing it into a combustion chamber, and sampling the emitted gases. We produced CWS and CWSP by the procedure from [38] in laboratory conditions (temperature 20 °C, pressure 101.3 kPa, relative humidity 80%). The prepared fuel sample was placed in a combustion chamber (muffle furnace) by means of a positioning mechanism. A modular probe collected samples of flue https://doi.org/10.1051/matecconf/201820900013

gases in the measuring area during fuel combustion in the chamber [13]. The gaseous products released during heating, ignition, and combustion of fuel in the muffle furnace were recorded and analyzed by the emissions monitoring system [13]. The mass of the sample was 1 g. Only a portion this small and a model combustion chamber [13] make it possible to meet the objectives of this research with the required degree of accuracy.

3 Results and discussion

Fig. 2 presents the concentrations of the main anthropogenic emissions (SO_x, NO_x) from the combustion of conventional coal dust and coal-water fuel (filter cake) versus coal-water slurries containing petrochemicals and additives. The figure clearly shows the differences between the main anthropogenic emissions when the additives under study are used.

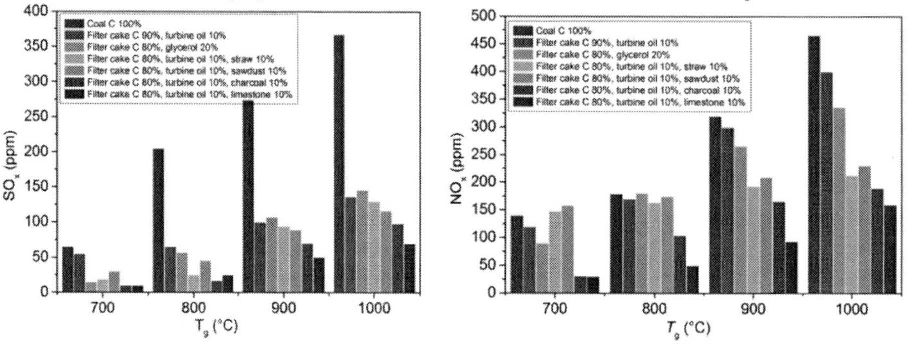

a *b*

Fig. 2. Concentrations of SO_x (*a*) and NO_x (*b*) emissions versus combustion temperature

The increased amount of water phase in the slurries reduced the concentrations of all the anthropogenic emissions under study by 10–40%. During the CWSP combustion, water (30–50%) serves as an oxidizer, which accelerates coal burnout. Using CWS makes power facilities more cost-effective by decreasing the underburning and reducing the contamination of heating surfaces in boiler units. An increase in water phase amount reduces the underburning (CO). Water in the form of hot vapor in the combustion zone provides a finer distribution of the carbon basis due to microexplosions of slurry droplets [15]. As a result, due to its more complete burnout, the amount of soot, benzapyrene, and secondary hydrocarbons in the waste gases is reduced. During the thermal dissociation of water, free molecules of oxygen and hydrogen are released: $H_2O=H_2+O_2$.

The oxygen formed during this reaction enhances combustion, whereas hydrogen and carbon monoxide act as reductants, decreasing the content of nitrogen and sulfur oxides in the flue gases: $NO_x+H_2=N_2+H_2O$; $NO_x+CO=N_2+CO_2$; $SO_x+H_2=S+H_2OSO_x+CO=S+CO_2$.

Water, as a component of composite liquid fuel, decreases the temperature and rate of adiabatic combustion, which reduces the anthropogenic emissions of NO_x and SO_x.

We consider pine sawdust, which is a wood industrial waste. The presence of 10% of sawdust in the CWSP provides a twofold reduction in the concentration of nitrogen oxides formed during fuel combustion (150–200 ppm vs. 350–650 ppm for CWSP based on a filter cake and turbine oil). The established decrease in the formation of fuel nitrogen oxides has several reasons. First, sawdust intensifies ignition (temperatures of thermal decomposition and ignition of sawdust are 200–300 °C lower than those of filter cakes) and increases the yield of carbon monoxide, which takes part in the reducing reactions forming free N_2 ($NO_x+CO=N_2+CO_2$. Second, the Fe content in sawdust leads to more reactions of NO_x reduction and N_2 formation ($2Fe+3NO=3/2\ N_2+Fe_2O_3$). Thus, in terms of NOx emissions, the priority of slurries containing 10% of sawdust becomes obvious. The emission of sulfur dioxides from the combustion of CWSP and CWSP including sawdust is comparable and does not exceed 100 ppm.

The experiments have revealed that the addition of charcoal may have a significant effect on the fuel ecology. Therefore, its use as an additive requires rational consumption. For instance, in order to reach the optimal environmental performance (Fig. 2), it is necessary to add 10% of charcoal to the fuel based on a filter cake and used turbine oil. Such a concentration makes it possible to reduce the main anthropogenic emissions (NO_x and SO_2) with a negligible rise in the cost of the initial fuel but a significant improvement to the energy performance.

The experimental data show that adding straw to CWSP significantly reduces gas emissions (Figs. 2). Thermochemical conversion (combustion) releases compounds of alkaline and alkaline-earth metals, which straw wastes contain in abundance. These compounds react with sulfur oxides (e.g., $2CaO+2SO_2+O_2=2CaSO_4$) to form substances that remain in the coal ash and contribute to sulfur retention reactions due to gaseous sulfur trioxide absorption. The products of such reactions are KCl and K2SO4, NaCl and Na2SO4, and a small amount of Ca and Mg-sulfates, which form most of the residue on thermally loaded surfaces in boiler units. On the other hand, due to high content of alkaline metals, the melting temperature of straw ash is relatively low. Consequently, there may be slag deposits on low-temperature surfaces. Moreover, high content of chlorine in the straw may aggravate the corrosion of boiler elements.

The curves in Fig. 2 show the concentrations of the main anthropogenic emissions for a slurry with glycerol. They are overall similar to those obtained for straw, sawdust, and charcoal. SO_4 and NO_x concentrations are much lower, since glycerol contains virtually no sulfur or nitrogen. Low adiabatic temperatures of the glycerol flame are yet another reason for a reduction in the so-called fuel nitrogen oxides. An important factor affecting the emission of sulfur oxides is the content of sodium salts in glycerol.

The analysis of flue gases has shown (Figs. 2) that the average concentrations of sulfur and nitrogen oxides throughout the temperature range of 700 to 1,000 °C are lower than for all the other compositions under study. The decrease in the emissions when using lime results from a reduction in the natural sulfur and nitrogen content. Lime compensates for the liquid fuel component, which makes a massive contribution to the formation of the main fuel oxides

4 Conclusions

This is a comprehensive research into the environmental and energy benefits and drawbacks of a large-scale use of CWSP based on industrial wastes and coals of different ranks. Using CWSP as fuel is an effective way to recover industrial wastes, in particular, used oils and coal processing wastes. Anthropogenic emissions from the combustion of waste-based CWSP are no higher than those from the conventional coal dust combustion.

This work was supported by the Russian Foundation for Basic Research (project 18-43-700001) and project VIU-ISHFVP-184/2018.

References

1. J. Xu, Y. Yang, Y.-W. Li, Fuel. **152,** 122 (2015)
2. F.-Q. Su, K.-I. Itakura, G. Deguchi, K.Ohga, Appl. Energy. **189**, 142 (2017)
3. F. Liu, T. Lyu, L. Pan, F. Wang, Energy Policy. **105**, 398 (2017)
4. *World Energy Outlook Special Report 2016: Energy and Air (*International Energy Agency, New York, 2016)

5. M. Rahman, D. Pudasainee, R. Gupta, Fuel Process. Technol. **158,** 35 (2014)
6. B. Hampf, K.L. Rodseth, **50**, 140 (2015)
7. S. Li, T. Xu, P. Sun, Q. Zhou, H. Tan, S. Hui, Fuel. **87**, 723 (2008)
8. H.M. Ertugrul, M. Cetin, F. Seker, E. Dogan, Ecol. Indic. **67**, 543 (2016) 543–555.
9. I.G. Nodelman, S.V. Pisupati, S.F. Miller, A.W. Scaroni, J. Hazard. Mater. **74**, 47 (2000)
10. J. Liu, X. Jiang, L. Zhou, H. Wang, X. Han, J. Hazard. Mater. **167**, 817 (2009)
11. D.O. Glushkov, P.A. Strizhak, M.Yu. Chernetskii, Therm. Eng. **63**, 707 (2016) G.S. Khodakov, Therm. Eng. 54, **36** (2007)
12. M.A. Dmitrienko, G.S. Nyashina, P.A. Strizhak, J. Hazard. Mater. **338**, 148 (2017)

A geometric approach to the modelling of critical phenomena for a spray combustion model

Elena Shchepakina[1,*]
[1]Samara National Research University, 34, Moskovskoye shosse, Samara 443086, Russia

Abstract. The paper is devoted to the modelling of the critical phenomena in multiscale combustion models. Such models are usually described by singularly perturbed systems of differential equations to reflect the significant distinction in characteristic relaxation times of different physicochemical processes. The paper proposes an approach for modelling of critical phenomena on the basis of the geometric asymptotic method of invariant manifolds. The critical phenomenon means as a sharp change in the dynamics of the process under consideration. As an illustration of this approach a dynamic model of fuel spray ignition and combustion is considered. The realizability conditions for the critical regime is obtained in the form of the asymptotic expression for the control parameter. The main feature of the critical regime is that during it the temperature of the combustible mixture can reach a high value within the framework of a safe process. It is shown that the critical regime plays the role of a watershed between the slow combustion regimes and the thermal explosion.

1 Introduction

A wide spectrum of thermophysical and chemical processes occurring during combustion, especially in multiphase media, is characterized by a significant difference in time scales (for example, temperature changes vs. concentration changes). Therefore, the singularly perturbed systems of differential equations are used to model combustion processes. For studying such systems different asymptotic methods are usually applied. The asymptotic geometric method used in this paper is based on the theory of invariant manifolds (see, for instance, [1] and the references therein). This method allows us to replaces the original system by another system on an invariant manifold of lower dimension. The lowering of the dimension occurs due to the decomposition of the original system in the vicinity of the integral surface into the independent "slow" subsystem and the "fast" subsystem. If the slow invariant manifold is attractive, then the analysis of the original system can be replaced by the analysis of the slow subsystem retaining a desired degree of accuracy. The use of such approach makes it possible not only to substantially simplify the initial system by reducing its dimension and by eliminating the computational rigidity, but it also gives us a full picture of the dynamics of the physical processes and provides an effective tool for

* Corresponding author: shchepakina@yahoo.com

modelling the critical phenomena in these processes. The critical regime corresponds to a chemical reaction separating the domains of self-accelerating reactions and domains of safe reactions. This regime is modelled by a unique trajectory that contains an unstable invariant manifold.

2 Spray ignition model

The model of the ignition and combustion of a fuel spray is suggested using an adiabatic approach [2-5]. The fuel spray is considered as a two-phase medium: combustible gas mixture — combustible liquid droplets. The pressure change in the reaction volume and its effect on the combustion process is assumed a slight and is omitted. The usual assumption is made that the thermal conductivity of the liquid phase is much greater than that in the gas phase. Thus, the heat transfer coefficient in the liquid gas mixture is supposed to be defined by the thermal properties of the gas phase. The drop boundary is assumed to be on a saturation line, i.e., the liquid temperature is constant and is equal to the liquid saturation temperature. The combustion reaction is modelled as a first order, highly exothermic chemical reaction. The model is built with the usual assumptions of the theory of combustion processes in chemical homogeneity at each point of the reaction vessel and in the dimensionless form has the form [4]:

$$\gamma\frac{d\theta}{d\tau}=\eta\exp\left(\frac{\theta}{1+\beta\theta}\right)-\varepsilon_1 r\theta(1+\beta\theta), \tag{1}$$

$$\frac{dr^3}{d\tau}=-\varepsilon_1\varepsilon_2 r\theta, \tag{2}$$

$$\frac{d\eta}{d\tau}=-\eta\frac{1}{1+\beta\theta}\exp\left(\frac{\theta}{1+\beta\theta}\right)+\varepsilon_1\psi r\theta, \tag{3}$$

where θ is the dimensionless fuel gas temperature; r is the dimensionless radius of the drops; η is the dimensionless concentration of flammable gas; τ is the dimensionless time; γ is the dimensionless parameter equal to the final dimensionless adiabatic temperature thermally isolated system after the explosion; β gives the initial temperature; ε_1,ε_2 characterize the interaction between the gas and liquid phases; ψ is a parameter characterizing the ratio of the energy of combustion gas mixture to the liquid evaporation energy.

The initial conditions for the equations (1)-(3) are:

$$\theta(0)=0, r(0)=1, \eta(0)=1.$$

Appropriate combination of equations (1)-(3) and integration over time yields the following energy integral

$$\eta-1+\frac{\gamma}{\beta}\ln(1+\beta\theta)+\frac{\psi-1}{\varepsilon_2}(r^3-1)=0,$$

which allows to reduce the order of the system (1)-(3) to the form

$$\gamma\frac{d\theta}{d\tau}=\left[1-\frac{\gamma}{\beta}\ln(1+\beta\theta)-\frac{\psi-1}{\varepsilon_2}(r^3-1)\right]\exp\left(\frac{\theta}{1+\beta\theta}\right)-\varepsilon_1 r\theta(1+\beta\theta) \tag{4}$$

$$\frac{dr^3}{d\tau}=-\varepsilon_1\varepsilon_2 r\theta. \tag{5}$$

Thus, the dynamic of the system depends on five dimensionless parameters: $\beta \ll 1, \gamma \ll 1, \varepsilon_1, \varepsilon_2, \psi$. The degenerate equation, which follows from the fast subsystem (4) for $\gamma = 0$, describes a slow curve

$$\Lambda(\theta, r) = \left[1 - \frac{\gamma}{\beta}\ln(1 + \beta\theta) + \frac{(\psi - 1)(1 - r^3)}{\varepsilon_2}\right]\exp\left(\frac{\theta}{1 + \beta\theta}\right) - \varepsilon_1 r\theta(1 + \beta\theta) = 0$$

in the phase plane. The flow of the system (4), (5) near the slow curve has a velocity of order of one as $\gamma \to 0$ while far from the slow curve the variable θ is changed very fast. In $\gamma -$ neighborhood of the slow curve there exists a slow invariant manifold of the system which is defined as an invariant surface of slow motions.

The set of points on the slow curve at which $\partial\Lambda/\partial\theta < 0$ ($\partial\Lambda/\partial\theta > 0$) forms the stable (unstable) part of the slow curve. The stable and unstable parts of the slow curve are the zero-order approximations ($\gamma = 0$) of the stable (or attractive) and unstable (or repulsive) slow invariant manifolds of the system (4), (5), respectively.

For $0 < \psi < 1 - \varepsilon_2$ the slow curve has the shape similar to that shown in Fig. 1. The part *PT* of the slow curve is stable while the part *TQ* is unstable. The ordinate of the point *T* depending on the parameters values can be equal to 1 or be greater (or less) than 1. If the point *T* has an ordinate greater than 1 then a trajectory of the system starting from the initial point tends to stable part *PT* of the slow curve and follows it to the origin. This is the case of the slow combustion regime; see the trajectory *CTP* in Figure 1.

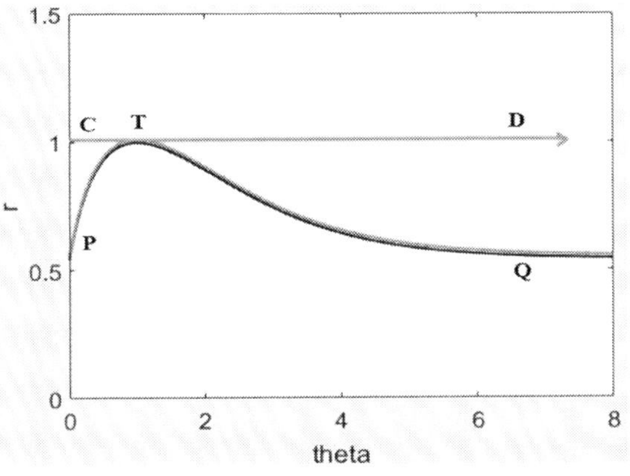

Fig. 1. Slow curve (black) and the trajectories (green) of (4), (5) in limit case ($\gamma = 0$)

If the point *T* has an ordinate less than 1 then a trajectory of the system starting from the initial point will pass beyond the basin of attraction of the stable part of the slow curve. This is the case of the explosion regime; see the trajectory *CD* in Figure 1.

The critical regime corresponds to the case when the trajectory of the system falls into a small vicinity of the point *T* and passes along the unstable part *TQ* of the slow curve; see the trajectory *CTQ* in Figure 1. The critical regime separates the explosive regimes from the safe combustion modes [1, 6]. The crucial result is that the unstable slow manifold may be used to construct the critical trajectory *CTQ* and to calculate the corresponding value of a control parameter, say, ε_1, of the system. For this goal we use the asymptotics proposed in [7].

The part *CT* of the critical trajectory can be represented in the form:

$$r(\theta, \gamma) = r^* + \gamma^{\frac{2}{3}}\Gamma_0^{\frac{2}{3}}\Omega\,\mathrm{sign}f(r^*, \theta^*) + \frac{1}{3}\gamma\ln\frac{1}{\gamma}\Gamma_1\mathrm{sign}f(r^*, \theta^*) + O(\gamma), \qquad (6)$$

where r^*, θ^* denote the coordinates of the point *T*,

$$g(\theta,r)=\left[-1+\frac{\gamma}{\beta}\ln(1+\beta\theta)+\frac{\psi-1}{\varepsilon_2}(r^3-1)\right]\exp\left(\frac{\theta}{1+\beta\theta}\right)+\varepsilon_1 r\theta(1+\beta\theta),$$

$$f(\theta,r)=\varepsilon_1\varepsilon_2 r\theta,\quad \omega=2.338107,\ \Gamma_0=\sqrt{\frac{2}{|g_{\theta\theta}(r^*,\theta^*)g_r(r^*,\theta^*)|}}|f(r^*,\theta^*)|,$$

$$\Gamma_1=\frac{6g_{\theta\theta}(r^*,\theta^*)f_\theta(r^*,\theta^*)-2g_{\theta\theta\theta}(r^*,\theta^*)f(r^*,\theta^*)}{3g_{\theta\theta}^2(r^*,\theta^*)}.$$

The functions f and g in the representation (6) are the right parts of the system (4), (5) after the transition to the inverse time that make the part TQ of the slow curve stable. The coordinates r^*,θ^* can be calculated from the system $g(r^*,\theta^*)=g_\theta(r^*,\theta^*)=0$.

Substituting all the found values into (6) and setting $r=1$ since the point C has the coordinate $r=1$, we obtain the equation for calculation of the critical values of the parameter $\varepsilon_1=\varepsilon_1^*$ in the form of asymptotic representation

$$\varepsilon_1^*=\varepsilon_{10}+\gamma^{\frac{2}{3}}\varepsilon_{11}+\gamma\ln\frac{1}{\gamma}\varepsilon_{12}+O(\gamma).$$

The direct calculation gives

$$\varepsilon_{10}=\frac{pe(\psi-1)}{\varepsilon_2},\ \varepsilon_{11}=-54\Omega(\psi-1)^3\sqrt[3]{\frac{2}{9}}\mu_1^{\frac{4}{3}}\mu_{23}^{-\frac{2}{3}}\varepsilon_{10}^{-\frac{4}{3}}\varepsilon_2^{-\frac{4}{3}}$$

$$\times\left\{(1-6\beta)\left[\frac{\psi-1}{\varepsilon_2}(\mu_{23}^3-1)-1\right]e+2\varepsilon_{10}\mu_{23}\beta\right\}^{-\frac{1}{3}}\left[3\mu_{23}^2e^{\frac{\psi-1}{\varepsilon_2}}+\varepsilon_{10}(1+\beta)\right]^{-\frac{1}{3}},$$

$$\varepsilon_{12}=\frac{\varepsilon_{10}\varepsilon_2}{3\mu_{23}}\left\{(1-14\beta)\left[\frac{\psi-1}{\varepsilon_2}(\mu_{23}^3-1)-1\right]e+4\varepsilon_{10}\mu_{23}\beta\right\}$$

$$\times\left\{(1-6\beta)\left[\frac{\psi-1}{\varepsilon_2}(\mu_{23}^3-1)-1\right]e+2\varepsilon_{10}\mu_{23}\beta\right\}^{-2},$$

where

$$q=\frac{\varepsilon_2+\psi-1}{\psi-1},\ \mu_1=\sqrt{\frac{q^2}{4}+\frac{\varepsilon_{10}^3\varepsilon_2^3}{27(\psi-1)^3}},\ \mu_2=\sqrt[3]{\frac{q}{2}+\mu_1},\ \mu_3=\sqrt[3]{\frac{q}{2}-\mu_1},\ \mu_{23}=\mu_2+\mu_3,$$

and p is a root of the equation

$$1=\sqrt[3]{\frac{q}{2}+\sqrt{\frac{q^2}{4}+\frac{p^3}{27}}}+\sqrt[3]{\frac{q}{2}-\sqrt{\frac{q^2}{4}+\frac{p^3}{27}}}.$$

The value $\varepsilon_1=\varepsilon_1^*$, at which the trajectory of (4), (5) contains the unstable manifold is said to be critical, i.e. this regime is not a slow combustion regime, since the temperature achieves a high value, and is not explosive, as the temperature increases at the tempo of the slow variable as this is the slow manifold. The value of $\varepsilon_1=\varepsilon_1^*$, gives the critical trajectory which separates the explosion modes ($\varepsilon_1<\varepsilon_1^*$) from the slow regimes ($\varepsilon_1>\varepsilon_1^*$) which are characterized by a slowdown of the reaction with small degrees of conversion and heating up is limited from above by $\theta<\theta^*$ at T.

Figure 2 shows the trajectory of the system (4), (5) and the solutions of the system (1)-(3) in the case of critical regime for $\varepsilon_1 = 2.2100108$, $\gamma = 0.01$, $\varepsilon_1 = 3.5, \varepsilon_2 = 0.8, \beta = 0.05, \psi = 0.19$.

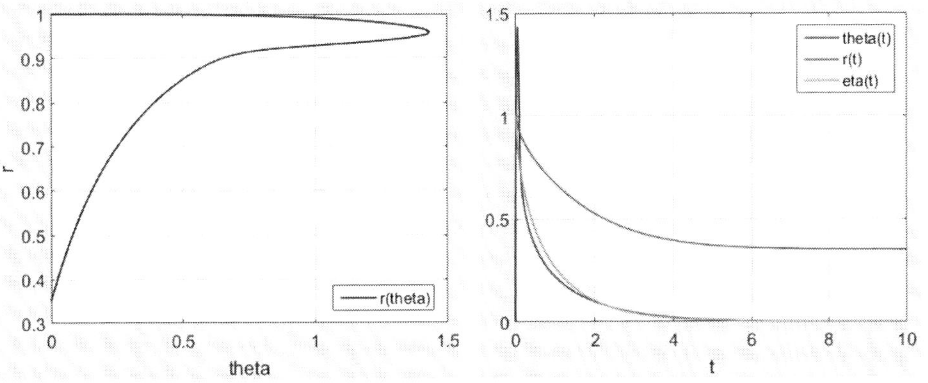

Fig. 2. Trajectory of (4), (5) and the solutions of (1)-(3) in the case of critical regime.

3 Conclusions

The dynamic model of a fuel spray ignition was studied by the methods of the geometric theory of singular perturbations. The realizability conditions for the critical regime was obtained in the form of the asymptotic expression for the control parameter. The main feature of the critical regime is that during it the temperature of the combustible mixture can reach a high value within the framework of a safe process. It was shown that the critical regime plays the role of a watershed between the slow combustion regimes and the thermal explosion.

The analytical calculations presented in this paper are of utmost importance for the realization of safe high-temperature processes.

This work was supported by the Ministry of Education and Science of the Russian Federation as part of a Program "Research and Development in Priority Areas of Advancement of the Russian Scientific and Technological Complex for 2014-2020" (RFMEFI58716X0033).

References

1. E. Shchepakina, V. Sobolev, M.P. Mortell, *Singular Perturbations. Introduction to System Order Reduction Methods with Applications* (Springer, Cham, 2014)
2. R. Stone, *Introduction to Internal Combustion Engines* (The MacMillan Press Ltd, 2nd Ed., 1992)
3. S. Sazhin, *Droplets and Sprays* (Springer, London, 2014)
4. I. Goldfarb, V. Gol'dshtein, I. Shreiber, A. Zinoviev, *Proceedings of the 26th Symposium on Combustion,* 1557 (The Combustion Institute, Pittsburgh, 1996)
5. I. Goldfarb, V. Gol'dshtein, G. Kuzmenko, S. Sazhin, Combust. Theory Modell. **3**, 769 (1999)
6. V.A. Sobolev, E.A. Shchepakina, Combust. Explos. Shock Waves **29**(3) 378 (1993)
7. E.F. Mishchenko, N. Kh. Rozov, *Differential Equations with Small Parameters and Relaxation Oscillations* (Plenum Press, New York, 1980)

MATEC Web of Conferences **209**, 00015 (2018) https://doi.org/10.1051/matecconf/201820900015
ComPhysChem'18

Numerical study of H_2S-H_2O-air mixture conversion to hydrogen via activation of air by an electric discharge

Ilya Arsentiev, *Vera* Savelieva, and *Nataliya* Titova*

Central Institute of Aviation Motors, Aviamotornaya 2, Moscow, 111116 Russia

Abstract. The numerical analysis of H_2 production during partial oxidation of H_2S–H_2O–air in a plug-flow reactor at a rather low temperature (T_0=500 K) was conducted. For the reforming process promotion, the oxidizer (air) was activated by an electrical discharge with different values of reduced electric field E/N and input energy E_s. It was shown that a significant hydrogen yield in a flow reactor can be obtained only after mixture ignition. The ignition delay length turned out to be minimal at E/N~4–10 and 120–150 Td, when $O_2(a^1\Delta_g)$ mole fraction in the discharge products is maximal. If the H_2S–H_2O–air mixture ignites inside the flow reactor, the H_2 mole fraction and its relative yield do not depend on E/N. The relative hydrogen yield increases monotonically with an increase of H_2O amount. The specific energy requirement for H_2 production in considered process was evaluated.

1 Introduction

Hydrogen sulfide is a part of associated petroleum and acid natural gases. It is also formed from the sulfur-containing fuels in petroleum refining industries. H_2S is a toxic and environmentally hazardous compound. Therefore, up to now, issues concerning the development of methods for its utilization remain very topical. The promising way of H_2S utilization is the H_2 production during H_2S partial oxidation [1-3]. It was shown earlier that the addition of water to H_2S-O_2(air) mixture allowed increasing the H_2 yield [3]. However, the conversion of H_2S to H_2 during partial oxidation of H_2S occurs only after H_2S ignition [3,4]. In order to ignite the fuel-rich H_2S–air mixture at a short residence time in the reactor, especially in the case with an admixture of water vapor, a significant gas heating is required. Plasma technologies can be used to enhance the ignition at a lower temperature [5], that allows decreasing the specific energy requirement and make the reforming process being more cost-effective. In this work, we analyze numerically the possibility of H_2 production during partial oxidation of H_2S in an atmospheric-pressure plug-flow reactor at a low initial temperature (T_0=500 K) of the H_2S-air mixture, when the air is preliminarily activated by an electrical discharge with different E/N values. The possibility of increasing the hydrogen yield in such a system via the addition of water vapor to H_2S is also investigated.

* Corresponding author: titova@ciam.ru

2 Thermodynamic analysis

For the determination of the optimal composition of the initial mixture from the point of view of maximal H_2 yield, we carried out thermodynamic calculations for H_2S–air initial mixture at T_0=500 K, P_0=1 atm and different values of fuel-to-air equivalence ratio ϕ under the conditions of constant pressure and enthalpy (Fig. 1). Maximal equilibrium mole fraction of H_2 turned out to equal to ~5% at ϕ=3–6. A Significant amount of H_2O, S_2, sulfur oxides (SO and SO_2), and H_2S can present in conversion products. Their mole fractions depend on ϕ value. At ϕ=2–3 there are mostly H_2O, SO_2 and S_2 in the equilibrium products. At $\phi \geq 4$ the main products are H_2O, S_2, and unreacted H_2S. Thermodynamic calculations have shown that mole fractions of S_3-S_8 species rise with the increase in ϕ value. But even at ϕ=10, their total amount does not exceed 0.15%. A small amount of solid sulfur precursors is due to the fairly high temperature T_e>1100 K of the conversion products when oxygen presents in the initial mixture.

Since the amount of H_2S in the initial mixture varies with the change of ϕ value, we will consider the relative amount of hydrogen molecules obtained from H_2S, which was determined as $\delta = n_{H_2} / n^0_{H_2S}$, where n_{H_2} and $n^0_{H_2S}$ are the amounts of H_2 moles in the conversion products and H_2S moles in the initial mixture, respectively (for flow reactor calculations, n_{H_2} and $n^0_{H_2S}$ denote corresponding molar flow rate of H_2 and H_2S). One can see from Fig. 1, that maximal equilibrium relative hydrogen yield at T_0=500 K is observed at ϕ=1.5–2 and equals to $(\delta^e)_{max}$~0.2. Note that maximal value of $(\delta^e)_{max}$ is obtained at the lower ϕ value, than that of $(\gamma^e_{H_2})_{max}$. Because the value $(\delta^e)_{max}$ is realized at ϕ~2, precisely such a value of ϕ was chosen to study the conversion process in the flow reactor. The addition of water vapor can increase δ^e (see Fig. 1). At water fraction $\varepsilon = n^0_{H_2O} / n^0_{H_2S} = 1$, where $n^0_{H_2O}$ is the amount of H_2O moles in the initial mixture, $(\delta^e)_{max}$ increases by 10% at ϕ=2 and by 25% at ϕ=1.5. $(\delta^e)_{max}$ is achieved at a lower ϕ value when adding H_2O than that in pure H_2S–air mixture. At ϕ=3, a quite large additive of water vapor (ε=1) can even reduce the relative H_2 yield. This is due to an essential decrease in temperature in the conversion products in the case of H_2O addition.

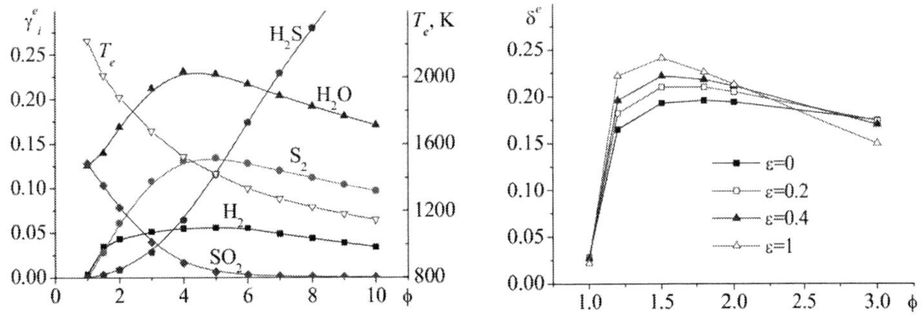

Fig. 1 – Equilibrium temperature (T_e), mole fractions of main species (γ^e_i) and relative H_2 yield (δ^e) as a function of ϕ value calculated for the initial H_2S–air mixture.

3 Methodology

Numerical analysis of the H_2 production during partial oxidation of H_2S was conducted in an atmospheric-pressure plug-flow reactor. The air, activated in an electrical discharge and passed through a postdischarge reactor, and H_2S–H_2O mixture are supplied separately to mixing reactor, and the homogeneous mixture proceeds to the flow reactor. Both flows are preliminarily heated up to the temperature T_0=500 K. An adiabatic plug-flow reactor was 1 m in length. The mixture velocity at the flow reactor inlet was U_0=1 m/s that corresponds to the gas residence time of $\tau_r \approx 1$ s. Calculations were carried out with the use of CHEMKIN software package.

The temperature and gas composition of air plasma generated by the electric discharge were calculated according to the ideal discharge model [6] using the Boltsig+ code [7] to solve the electron Boltzmann equation. The model of air discharge includes 40 processes with electrons: elastic collisions, excitation of rotational, vibrational and electronic states of O_2 and N_2 molecules, dissociation and ionization of O_2 and N_2 molecules. The dependencies of processes cross sections on electron energy were taken from [8] in form of tables, suitable for the Boltsig+ code. Secondary plasma-chemical reactions, as well as quenching of electronically excited species and vibrational relaxation, were taken into account during passing the discharge products through a postdischarge channel with the residence time of 1 ms. Mechanism of plasma-chemical reactions was borrowed from [9].

One can see from Fig. 2 that, to the postdischarge reactor outlet, an essential part of the discharge energy transfers to heating the air. The air plasma temperature is in the range 880–893 K at E/N=1–150 Td and E_s=0.5 J/ncm^3. Besides the initial species, $O_2(a^1\Delta_g)$ molecules are most abundant. Their maximal mole fraction is ~0.3% at E/N~6 Td and E_s=0.5 J/ncm^3. $O_2(a^1\Delta_g)$ concentration has a dip around E/N~40 Td. O atoms are formed at higher values of E/N>40 Td. Note that, at E/N~60–70 Td, mole fractions of $O_2(a^1\Delta_g)$ molecules, and O atoms are practically the same. Concentrations of other excited species, except $O_2(b^1\Sigma_g^+)$, and ions are very low. This fortunate fact allows us to include only $O_2(a^1\Delta_g)$ and $O_2(b^1\Sigma_g^+)$ species into the kinetic model of H_2S oxidation.

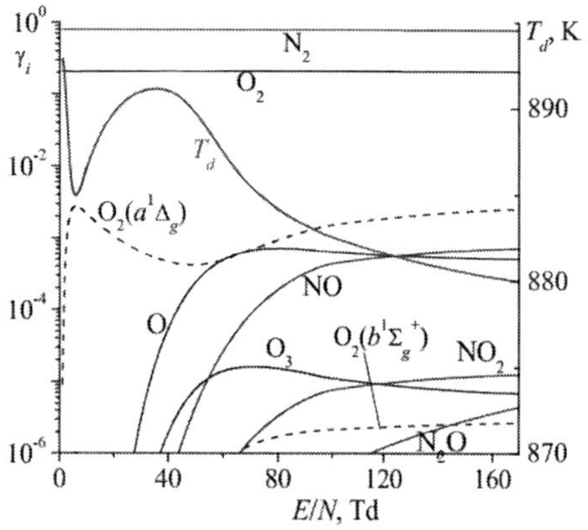

Fig. 2. Gas temperature T_d and mole fractions γ_i of the main species at postdischarge reactor outlet as a function of E/N value in the air discharge with E_s=0.5 J/ncm^3 at T_0=500 K.

For the numerical study of H_2S-H_2O-air mixture conversion in a flow reactor, the reaction mechanism [3] was taken as a basic one. This mechanism was supplemented by the reactions with participation of $O_2(a^1\Delta_g)$ and $O_2(b^1\Sigma_g^+)$ molecules which were taken from [10].

4 Results and discussion

Fig. 3 shows the dependencies of $L_{in}(E/N)$ for two values of the supplied energy E_s=0.3 and 0.5 J/ncm^3. One can see that at E_s=0.5 J/ncm^3 the H_2S–H_2O–air mixture ignites inside the reactor at any E/N value and ε=0–1 (L_{in}<1 m). But the shortest ignition length is observed for the discharge that produces more $O_2(a^1\Delta_g)$ molecules. This occurs at E/N~4–10 and 120–150 Td. Recall that, at E/N~60–70 Td, the mole fractions of O atoms and $O_2(a^1\Delta_g)$ molecules are equal (see Fig. 2). Analysis showed that the contribution of O atoms and $O_2(a^1\Delta_g)$ molecules to the ignition acceleration was almost the same. But since the total number of O and $O_2(a^1\Delta_g)$ in this case is two times smaller than the mole fraction of $O_2(a^1\Delta_g)$ molecules at E/N~6 Td, the ignition occurs later. At a lower energy input

E_s=0.3 J/ncm^3, the mixture ignites only at ε=0–0.2 and only at *E*/*N* values that provide a maximal number of $O_2(a^1\Delta_g)$ molecules.

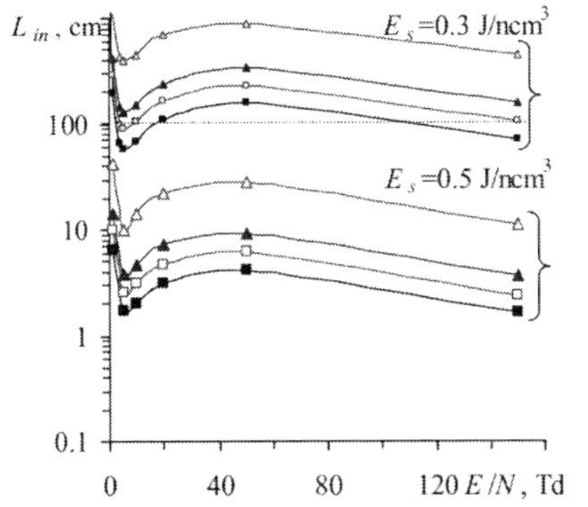

Fig. 3. Ignition delay length L_{in} depending on *E*/*N* during partial oxidation of H_2S–H_2O–air mixture with ϕ=2 and ε=0, 0.2, 0.4 and 1 in the flow reactor when air is exposed to the electric discharge with E_s=0.3 and 0.5 J/ncm^3, T_0=500 K, P_0=1 atm.

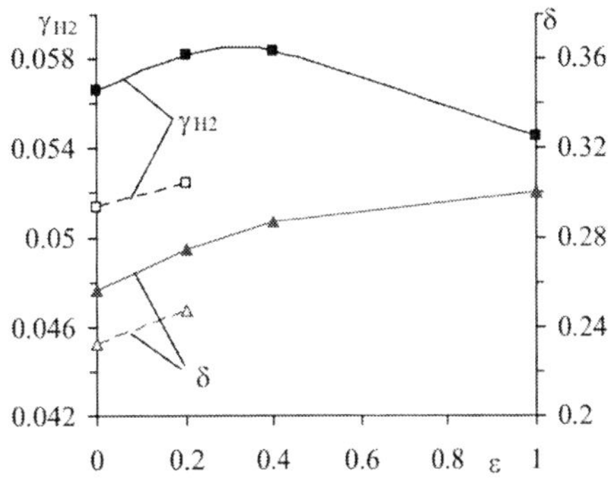

Fig. 4. Mole fraction γ_{H2} and relative yield δ of H_2 depending on water content ε during partial oxidation of the H_2S–H_2O–air mixture with ϕ=2 in the flow reactor when air is exposed to the electric discharge with E_s=0.3 and 0.5 J/ncm^3 (dashed and solid lines) at T_0=500 K, P_0=1 atm.

If the mixture ignites, quantitative H_2 yield does not depend on *E*/*N*. This occurs at E_s=0.5 J/ncm^3. At E_s=0.3 J/ncm^3 hydrogen can be produced in the flow reactor in two ranges of *E*/*N*~4–15 and 120–150 Td if ε=0, and only at *E*/*N*~4.5–8 if ε=0.2. If the mixture ignites, then a small addition of water vapor (ε=0–0.3) slightly increases H_2 mole fraction in the conversion products (Fig. 4). With further increase of H_2O amount, γ_{H_2} value decreases. The smaller H_2 yield is obtained at E_s=0.3 J/ncm^3, which is due to a decrease in the conversion products temperature. The relative H_2 yield, on the contrary, increases with an increase of H_2O in the mixture, from δ=0.25 at ε=0 to δ=0.3 at ε=1.

The main reaction responsible for the H_2 production is the process

$$H_2S+H=HS+H_2. \qquad \text{(R1)}$$

The formed H_2 is consumed before the ignition in the reaction

$$H_2O+H=OH+H_2 \qquad \text{(R2)}$$

that occurs in the backward direction with the formation of H_2O. However, just after the ignition event, the reaction (R2) starts to occur in the forward direction, resulting in the consumption of H_2O and formation of the additional amount of H_2. An admixture of H_2O additionally shifts the equilibrium of this reaction towards the formation of H_2, thus increasing hydrogen yield. The larger the amount of H_2O in the initial mixture, the greater the role of the reaction (R2). Fig. 5 shows the H_2 production rates for the case of ϕ=1.5 when, according to equilibrium calculations (Fig. 1), the most pronounced effect of H_2O addition on the increase of relative H_2 yield can be expected.

The addition of water vapor decreases the conversion degree of H_2S ($\alpha_{H_2S} = (n^0_{H_2S} - n_{H_2S})/n^0_{H_2S}$) from 97% at ε=0 to 93% at ε=1. Simultaneously, the relative yield of SO_2 increases from 0.36 to 0.42 due to the additional amount of oxygen in the system.

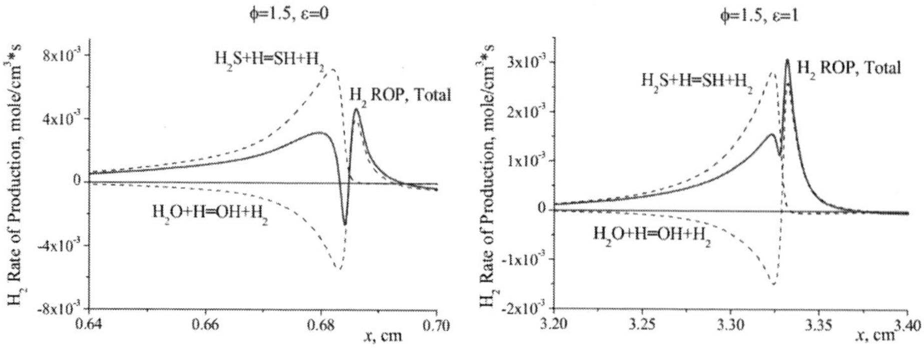

Fig. 5. Variation of H_2 production rates during partial oxidation of the H_2S–H_2O–air mixture in the flow reactor at ϕ=1.5 and ε=0 and 1 when air is exposed to the electric discharge with E_s=0.5 J/ncm³ at T_0=500 K, P_0=1 atm.

For the evaluation of the energy efficiency of the process considered in this paper, we will use the parameter characterizing the specific energy requirement for production of H_2 molecule: $\mathrm{SER}(\mathrm{H}_2) = (Q + E_s)/n_{\mathrm{H}_2}$, where Q is the energy required to heat the H_2S–H_2O and air flows from room temperature to 500 K. The minimum value of SER(H_2), obtained for the H_2S–air mixture at ϕ=2, E_s=0.3 J/ncm³ and E/N~10 Td, turned out equal to 10.5 eV/(molecule H_2). This value is essentially higher than the minimum SER(H_2)=0.83 eV/(molecule H_2) value for the H_2S–O_2 mixture at the same conditions, because, when using air as an oxidizer, a lot of energy is spent on heating and excitation of N_2 molecules, which do not participate in the conversion process. Therefore, energy is wasted inefficiently. The use of air for H_2S partial oxidation seems to be not so promising compared to the use of oxygen as an oxidizer. However, economic evaluation of the H_2S conversion process should also include the oxygen cost.

The work was supported by the Russian Science Foundation (project no. 16-19-00111) and by the Russian Foundation for Basic Research (project no. 17-01-00810-a) in the part of the analysis of chain branching in low-temperature H_2S oxidation when activating air in an electric discharge.

References

1. V. Palma, V. Vaiano, D. Barba, M. Colozzi, E. Palo, L. Barbato, et al., Int. J. Hydrogen Energy **40**, 106 (2015)
2. T.Yu. Cong, A. Raj, J. Chanaphet, S. Mohammed, S. Ibrahim, A.A. Shoaibi, Int. J. Hydrogen Energy 41, 6662 (2016)
3. V.A. Savelieva, N.S. Titova, A.M. Starik, Int. J. Hydrogen Energy **42**, 10854 (2017)
4. V.A. Savelieva, A.M. Starik, N.S. Titova, O.N. Favorskii, Combust., Expl., Shock Waves **54**, 136 (2018)
5. A.M. Starik, P.S. Kuleshov, B.I. Loukhovitski, N.S. Titova, Int. J. Hydrogen Energy **40**, 9872 (2015)
6. A.M. Starik, B.I. Loukhovitski, A.P. Chernukho, Plasma Sources Sci. Technol. **21**, 035015 (2012)
7. G.J.M. Hagelaar, L.C. Pitchford, Plasma Sources Sci. Technol. **14**, 722 (2005)
8. A.V. Phelps, http://jila.colorado.edu/~avp/collision_data/electronneutral/ ELECTRON.TXT
9. A.M. Starik, N.S. Titova, I.V. Arsentiev, Plasma Sources Sci. Technol. **19**, 015007 (2010)
10. A.M. Starik, V.A. Savelieva, A.S. Sharipov, N.S. Titova, Combust. Flame **170**, 124 (2016)

Ozone recovery in the presence of CO and N_2O

Aleksei Torbin[1,2,*], *Andrey* Pershin[1,2], *Marsel* Zagidullin[1,2], *Michael* Heaven[1,3], *Alexander* Mebel[1,4], and *Valeriy* Azyazov[1,2]

[1]Samara National Research University, 443086 Samara, Russia
[2]Lebedev Physical Institute, 443011 Samara, Russia
[3]Emory University, 30322 Atlanta, Georgia, USA
[4]Florida International University, 33199 SW 8th St, Miami, USA

Abstract. Extremely reactive vibrationally excited ozone $O_3(\upsilon)$ molecule is formed in excess of O_2 in particular in air-fuel mixture, however there is a lack of kinetic data on $O_3(\upsilon)$ reactions with combustion products. The time-resolved absorption spectroscopy method was applied for measurements of the O_3 recovery rates after 266 nm photolysis in the presence of O, CO and nitrous oxides. It was revealed that reaction $O_3(\upsilon)$ + O results in an incomplete ozone recovery. In the reaction $O_3(\upsilon)$ + O stabilization channel dominates the reactionary one. In the presence of N_2O, ozone does not restore to its initial level.

1 Introduction

Active oxygen species (AOS – ozone O_3, molecular singlet oxygen $O_2(a)$ (0.98 eV) and $O_2(b)$ (1.63 eV), atomic oxygen $O(^3P)$ and $O(^1D)$ (1.97 eV)) play an important role in the combustion, especially when it is initiated by plasma [1, 2]. In recent years, a wide interest to them is because the processes involving AOS can be used to expand the limits of combustion and ignition. In order to reduce harmful emissions, primarily nitrogen oxides NO_x, requires reduction of the maximum gas temperature in the combustion chamber. For this purpose, the use of lean air-fuel mixture is preferable, as this also leads to a reduction in fuel consumption. Expanding the limits of combustion and ignition with the increase in the ratio "air/fuel" is one of the most pressing problems in the theory of combustion. These limits can be extended by external impact: plasma, chemical or the light (laser).

Creation of non-equilibrium plasma with high levels of reduced electric field, E/N, and a significant level of energy load may provide a high yield of oxygen atoms at O_2 plasma dissociation. The oxygen atoms enter into rapid chemical reactions with hydrocarbons and cause initiation of the combustion process. The possibility of using an electric discharge has been experimentally demonstrated in terms of thermal engines, such as in [1, 3]. Selective excitation of internal degrees of freedom of ozone and oxygen molecules by laser radiation can also increase the limits of the ratio "air/fuel".

With an excess of O_2 (the atmosphere, the air-fuel mixture, oxygen-containing plasma), oxygen atoms are effectively removed in the three-body recombination process

$$O + O_2 + M \rightarrow O_3(\upsilon) + M. \tag{1}$$

* Corresponding author: torbinalex@gmail.com

This forms a vibrationally-excited ozone molecule $O_3(\upsilon)$ which, together with the atomic oxygen may play an important role in the initiation of combustion, recovery of ozone in the atmosphere, etc. These processes were not previously included in the kinetic scheme of planetary atmospheres, combustion and oxygen-containing plasma. The vibrational excitation of ozone greatly accelerates the rate of reactions with O and $O_2(a)$ [4]. To determine the role of the reaction pathways to the dynamics of ozone in the atmosphere, in the combustion and in oxygen-containing plasma, new kinetic data are required that are not currently available in the literature and databases.

2 Experimental approach

The ozone kinetics was studied using the absorption spectroscopy method [4]. Ozone photolysis was performed by UV laser pulses at a wavelength of 266 nm (radiation of the 4th harmonic of the Solar Laser Systems, model LQ829 solid-state laser with a pulse duration of 10 ns) in a photolysis cell.

Pulsed laser photolysis of O_3/O_2 mixtures, in the presence of various buffer gases, was used for these kinetic measurements. A schematic diagram of the apparatus used is shown in Fig. 1. The photolysis radiation at 266 nm was provided by the fourth harmonic of a Nd:YAG laser (Solar Systems LQ829; 10 ns pulse duration; repetition rate 10 Hz). The diameter of the laser beam was 8 mm. The photolysis cell constructed from a solid brass block with 1-cm diameter channels for the flow of gases, the passage of the photolysis and probe laser beams, and the observation of emission from the photolysis zone. The gas velocity in the photolysis zone was typically 50–100 cm/s, which was fast enough to ensure that a fresh sample of gas was photolyzed on each pulse. A mass flowmeter and needle valves were used to control the gas flow rates. The cell was evacuated by a rotary pump, and the pumping rate was adjusted using a ball valve.

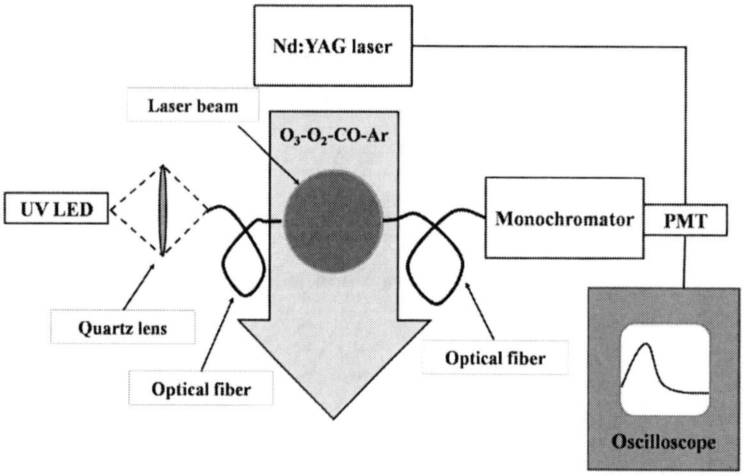

Fig. 1. Schematic diagram of the experimental setup for measuring the ozone concentration in the photolysis zone, based on time-resolved absorption spectroscopy.

Time-resolved absorption spectroscopy at λ=258 nm was employed for O_3 concentration measurements in the photolysis zone. A light-emitting diode (LED; UVTOP255) was used as a source of UV light. A quartz lens (f = 1.5 cm) was used to collect the light from the LED and focus this light into a quartz optical fiber. The light transmitted by this fiber traversed the photolysis zone, and a fraction of this light was collected by a second fiber. The ends of the optical fibers were positioned such that all of the O_3 between them was exposed to the photolysis radiation. The light transmitted by the collection fiber was dispersed by a monochromator and detected by a photomultiplier (PMT). Time-resolved emission signals from the PMT were averaged and stored using a digital oscilloscope (Agilent DSO1022A; 200 MHz response).

3 Results and analysis

3.1 Ozone recovery in the presence of O_2

Figure 2 shows the example of time dependences of ozone concentrations after laser photolysis of O_3-O_2-Ar mixture by radiation with a wavelength of 266 nm at the specific laser pulse energy E = 70 mJ/cm^2, total gas pressure P_{tot} = 712 Torr, and initial ozone pressure P_{O3} = 0.85 Torr. The gas mixture temperature was T = 300 K. The total mixture pressure and the total gas flow rate were maintained constant. Partial argon P_{Ar} and oxygen P_{O2} pressures were varied so that the equality $P_{O2} + P_{Ar} = P_{tot}$ would be satisfied.

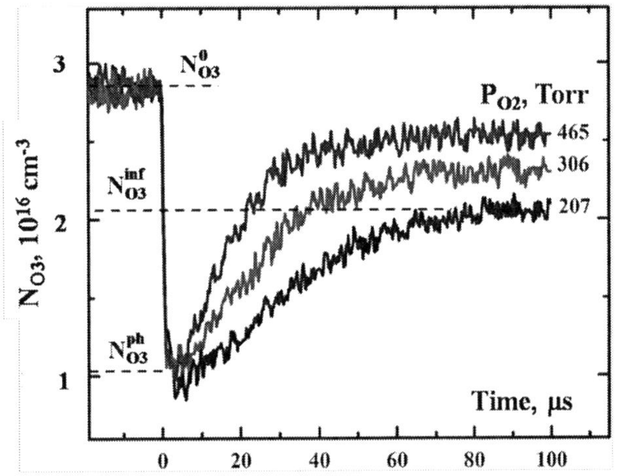

Fig. 2. Temporal profiles of the O_3 concentration at E = 70 mJ/cm^2, P_{tot} = 712 Torr, T = 300 K, and various O_2 pressures.

As seen in Fig. 2 the rate and degree of ozone recovery strongly depend on the mixture composition. We can see that the ozone concentration does not reach initial values N^0_{O3} and tend at $t \to \infty$ to certain values of N^{inf}_{O3}. This phenomenon was called the incomplete ozone recovery effect [5]. An addition of components quenching vibrationally excited ozone to the mixture increases the degree of ozone recovery, determined from the expression

$$\eta_{rec} = \frac{N^{inf}_{O3} - N^{ph}_{O3}}{N^{0}_{O3} - N^{ph}_{O3}},$$

where N^{ph}_{O3} – ozone concentration after laser photolysis. Ar is a less weak relaxant for $O_3(\upsilon)$ than O_2 , therefore, its addition to the mixture led to a decrease in the degree of O_3 recovery. The effect of incomplete ozone recovery is caused by the process $O_3(\upsilon)$+O with the branching factor $\gamma_r = 0.81 \pm 0.13$ for the chemical channel of reaction products.

3.2 Ozone recovery in the presence of CO

Figure 3 shows the typical time dependences of the ozone concentration after laser photolysis of mixture O_2-O_3-Ar-CO at a wavelength of 266 nm for specific pulse energy $E = 70$ mJ/cm^2, total gas pressure $P_{tot} = 720$ Torr, oxygen pressure $P_{O2} = 180$ Torr, gas mixture temperature $T = 300$ K, and initial ozone pressure $P_{O3} = 0.85$ Torr.

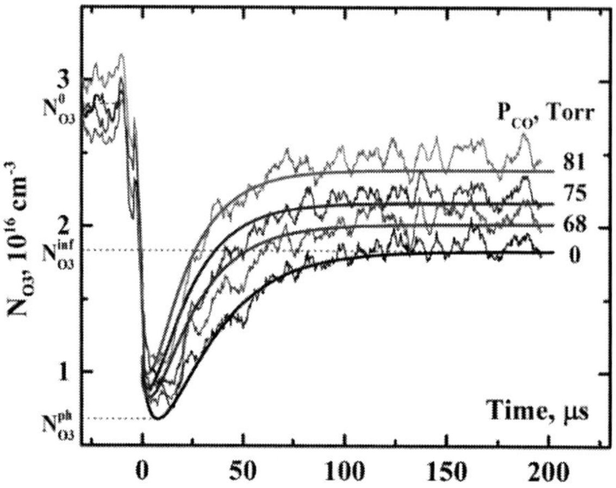

Fig. 3. Temporal profiles of the O_3 concentration at $E = 70$ mJ/cm^2, $P_{tot} = 720$ Torr, $P_{O2} = 180$ Torr, $T = 300$ K, and various CO pressures. Smooth curves are time profiles calculated by kinetic modeling.

The reaction of vibrationally excited ozone with carbon monoxide at moderate temperatures can proceed in two possible channels:

$$\text{reaction channel } O_3(\upsilon) + CO \rightarrow CO_2 + O_2, \quad (2)$$

$$\text{stabilization channel } O_3(\upsilon) + CO \rightarrow O_3 + CO. \quad (3)$$

The experimental results clearly show that an increase in the CO concentration leads to an increase in the degree of ozone recovery. This indicates that the stabilization channel dominates over the reaction one. Kinetic modeling of processes in the photolysis cell during experiments was performed. The best agreement with experimental data was achieved at the process (3) rate constant of $(1.5 \pm 0.2) \cdot 10^{-13}$ cm^3/s. The method used in this study does not allow an accurate conclusion about the contribution of the reaction channel to the general kinetics of changes in the ozone O_3 concentration, except for the estimate of the upper boundary of the reaction (2) rate constant not exceeding 10^{-14} cm^3/s.

3.3 Ozone recovery in the presence of nitrious oxides

Typical temporal profiles of the ozone concentration after laser photolysis of mixture O_3-O_2-N_2O at a wavelength of 266 nm for specific pulse energy $E = 70$ mJ/cm^2, total gas pressure $P_{tot} = 807$ Torr, oxygen pressure $P_{O2} = 600$ Torr, gas mixture temperature $T = 300$ K, and initial ozone pressure $P_{O3} = 0.85$ Torr are presented in Fig. 4.

Reaction of vibrationally excited ozone with nitrogen oxide NO produced in the secondary processes

$$O_3(\upsilon) + NO \rightarrow NO_2 + O_2 \quad (4)$$

has more probability then reaction with thermalized ozone with the rate constant 1.8×10^{-14} cm^3/s [6]. Reaction (4) competes with the relaxation process

$$O_3(\upsilon) + N_2O \rightarrow O_3 + N_2O \quad (5)$$

for which there is no any kinetic information.

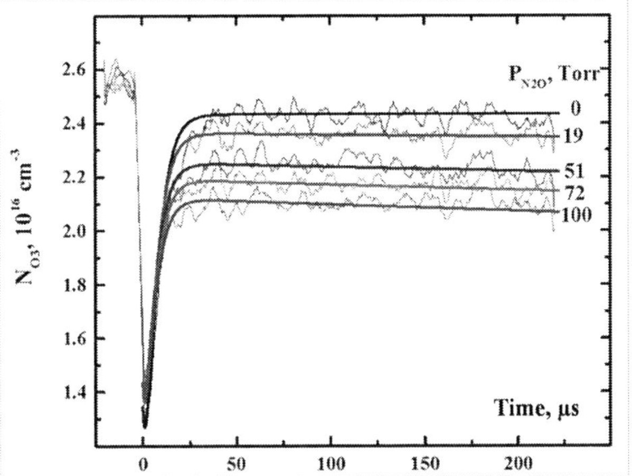

Fig. 4. Temporal profiles of the O_3 concentration at E = 70 mJ/cm^2, P_{tot} = 807 Torr, P_{O2} = 600 Torr, T = 300 K, and various N_2O pressures. Smooth curves are time profiles calculated by kinetic modeling.

The experimental results in Fig. 4 show that addition of N_2O into the gas mixture significantly decreases the degree of ozone recovery. Unfortunately the method used in this study does not allow to separate processes (4) and (5) because NO is actively produced due to reactions of N_2O with photolysis products. Thus we used kinetic modeling to estimate the orders of magnitude of the rate constants for the processes (4) and (5) which amounted to ~10^{-11} cm^3/s and ~10^{-13} cm^3/s, respectively.

Summary

The vibrationally excited ozone molecule $O_3(\upsilon)$ plays an important role in oxygen containing mixtures, thus processes involving it should be included into combustion and atmosphere databases. The degree of ozone recovery depends on the medium composition. An addition of components quenching $O_3(\upsilon)$, such as carbon monoxide CO, to the gas mixture increases the degree of ozone recovery. The reverse situation is observed with increasing the N_2O pressure which indicates the course of reactions between $O_3(\upsilon)$ and nitrous oxides. Accurate measurements of the rate constants for processes (4) and (5) require further investigation.

The work was supported by the Ministry of Education and Science of the Russian Federation under the Grant No. 14.Y26.31.0020 to Samara University

References

1. T. Ombrello, S.H. Won, Y. Ju, S. Williams, Combustion and Flame, **157** (2010)
2. A.A. Konnov, Combustion and Flame, **162**, 10 (2015)
3. K.V. Savelkin, D.A. Yarantsev, I.V. Adamovich, S.B. Leonov, Combustion and Flame, **162**, 3 (2015)
4. V.N. Azyazov, M.C. Heaven, International Journal of Chemical Kinetics, **47**, 2 (2015)

5. A.A. Pershin., A.P. Torbin, M.C. Heaven, V.N. Azyazov, Bulletin of the Lebedev Physics Institute, **43**, №1 (2016)

6. IUPAC. URL: http://iupac.pole-ether.fr/ (accessed: 25.12.2017)

Stabilization of combustion front in supersonic flow using streamer's discharge

Pavel Bulat[1,2], *Lev* Grachev[3], *Igor* Esakov[3], *Vladimir* Upyrev[1,2*]

[1]Baltic State Technical University "Voenmeh", 190005 Saint-Petersburg, Russia
[2]Saint Petersburg National Research University of Information Technologies, Mechanics and Optics, 197101 Saint-Petersburg, Russia
[3] Moscow Radiotechnical Institute of the Russian Academy of Sciences, 117519 Moscow, Russia

Abstract. Thermodynamic analysis shows that for flights with velocities exceeding six sound velocities, it is required to burn fuel not in a subsonic but in a supersonic flow. The aim of this work is to investigate the possibility of creating a stationary combustion front in a supersonic flow by igniting the mixture with an attached microwave discharge. Discharges are created on the resonator by means of a pulsed source of quasi-optical microwave radiation. This method of initiation is one or two orders of magnitude more economical than other known methods of plasma ignition and combustion stabilization. A numerical evaluation and comparison with experiment of the propagation velocity of a subcritical streamer discharge in a stationary medium and in a supersonic drifting flow are performed. Experiments have been conducted to ignite a flat flow of propane-air mixture, as well as ignition of the propane stream fed into the airflow, which simulates the operation of the fuel injector. In all cases, the experiments confirmed a steady fuel combustion, which was controlled by the temperature measurements with a thermocouple.

1 Introduction

The issue of initiation and maintenance of combustion in ramjet air-jet engines, including engines with supersonic combustion, is considered. At the moment, such problems are solved by placing in the flow of poorly streamlined bodies and the stabilization of combustion in the circulation zones by such bodies. This leads to an additional resistance, which increases as the flight speed and the speed of the ignited fuel mixture increase. An alternative variant of ignition of a mixture by means of a plasma discharge, a streamer and an attached diffuse one, is considered in this work. Due to the use of the resonant method of amplifying the electric field, these discharges ignite at the initial field strength much lower than the breakdown energy E_{cr}. Detailed studies of such discharges carried out during the last 20 years at the RAS of the Russian Academy of Sciences have shown that the above-mentioned method is 20 to 80 times higher in terms of energy costs than other methods, including corona discharge and spark ignition [1, 2].

* Corresponding author: upyrevvv@yandex.ru

The paper studies the initiation of combustion of a plane premixed air and propane stream at different speeds and fuel-oxidant ratios, as well as ignition of the propane stream fed into the air flow, similar to the operation of the fuel injector. Also in the paper the semi-empirical and numerical modeling of subcritical streamer discharge and deeply subcritical attached diffuse discharge is considered.

2 Mathematical model

When the microwave field is turned on and when free electrons are present in this region, their number increases with time t, and the region occupied by the discharge plasma diffusely "blur". This process at the initial stage is described by the following equation for the electron density

$$\frac{\partial n_e}{\partial t} = D_e \cdot \frac{\partial^2 n_e}{\partial r^2} + \left(v_i - v_a\right) \cdot n_e, \quad (1)$$

where n_e – concentration of electrons, $D_e = 7500/p$ – electron diffusion coefficient, p – gas pressure in Pa, t – time, v_i – ionisation frequency, v_a – frequency of attachment of electrons to atoms, r – radius vector. In the focus of the microwave beam, in contrast to the constant field, the average electron velocity in the avalanche is zero. Proceeding from this, it is possible to obtain the propagation velocity of the plasmoid boundary [3]:

$$u_s = 2\sqrt{D_e \cdot \left(v_i - v_a\right)} + \frac{|\mu| \cdot E}{\omega + \left(v_i - v_a\right)}\left(v_i - v_a\right) \quad (2)$$

where the first term is determined by electron diffusion, and the second by their drift in an electric field along the tension vector E, μ - mobility of electrons, ω – circular microwave radiation frequency.

In a motionless gas the problem is simplified and it is possible to estimate the velocity of the streamer propagation, based on the following considerations. Since outside the diffuse discharge $E < E_{cr}$,then for the development of the streamer the density of the gas inside it must fall, otherwise deionization will occur and the channel will lose the ability to increase the field strength to $E > E_{cr}$. Studies [4] show that the level of supercriticality E / E_{cr}, realized at the streamer head, is set at a level slightly higher than one, at which the velocity of the ionization front corresponds to the rate of channel heating directly at the head of the streamer. Thus, the second term in Eq. (2) is negligibly small and it can be ignored when estimating the growth rate of a subcritical streamer. The expression for the head speed of the streamer

$$u_s = 2\sqrt{D_e \cdot \frac{E_{cr}^2}{c_p \cdot p} \cdot \frac{\omega}{4 \cdot \pi \cdot \ln\left(\frac{E_{cr}}{E}\right)}}, \quad (4)$$

where c_p – specific heat at constant pressure.

3 Streamer velocity

Figure 1 shows the results of calculating the speed and radius of the streamer in comparison with the experimental data obtained when shooting discharges with a high-speed Phantom v.251 camera at a speed of 510000 frames / s. Figure 2 illustrates the influence of supersonic air flow on a subcritical microwave discharge. The streamer microwave discharge practically does not change its shape, but its plasma channels are blurred.

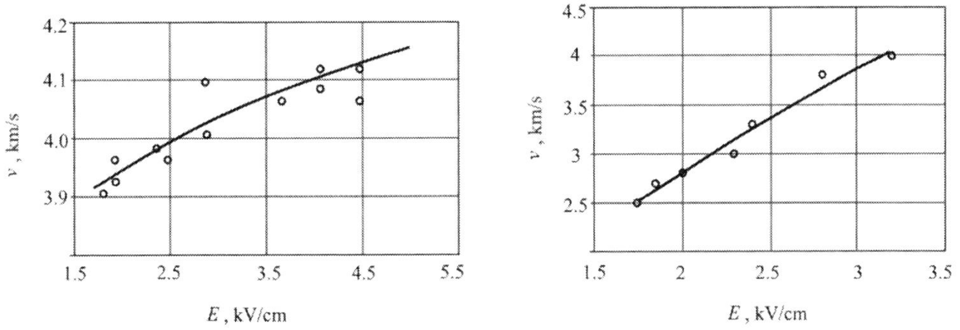

a) b)

Fig. 1. The propagation velocity of the streamer discharge in a front free space (a) and in a quartz tube (b) at atmospheric pressure, depending on the field strength in the region of the initiator: (—) – calculation by formula (4), (□ ◦) – the results of the experiment.

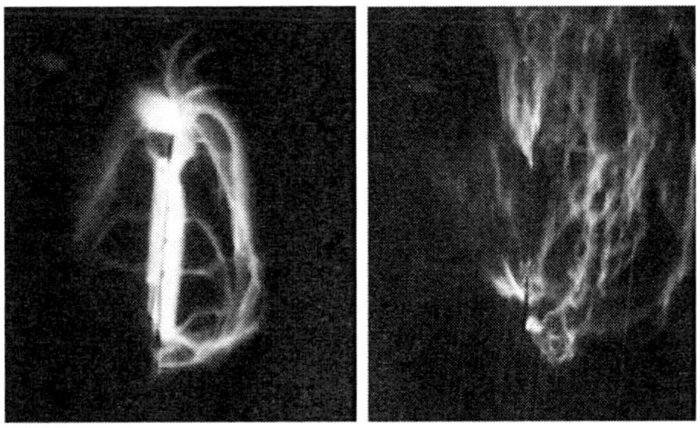

Fig. 2. Subcritical streamer discharge in still air (a) and in a supersonic flow at the velocity $V = 500$ m/s (b). Air pressure $p = 13.3$ kPa, field strength E_0=1.7 kV/cm.

4 Experimental setup

The scheme of the experimental setup is shown in Figure 3. Installation is a pulsed vacuum-type wind tunnel. The premixed propane mixture with air exits the nozzle at a speed of 10 to 500 m / s. The speed is controlled by the total pressure in the receiver. In the first case, a cylindrical vibrator with rounded ends was used to initiate combustion (Fig. 3a). In the second, a tubular initiator with a quartz head (Fig. 3b), through which propane was pumped. The tube itself was placed in the external air flow instead of the cylindrical initiator of the installation figure 3a. The thermocouple and the pitot tube were installed at the stagnation point.

 https://doi.org/10.1051/matecconf/201820900017

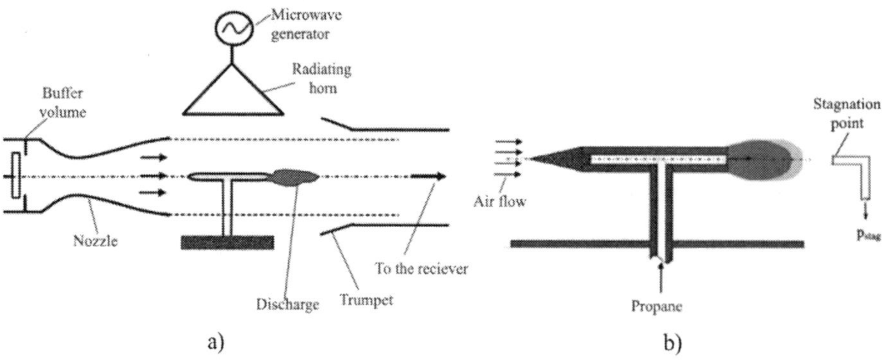

Fig. 3. The scheme of the experimental setup.

5 Results

The experimental results for the premixed fuel mixture are shown in Fig. 4. Using the cap, the flow is flat. Its speed is 30 m / s. The temperature at the stagnation points is shown to the right. The flame is visible even for the case of a particularly poor mixture (Fig. 4e). In [4], we have already investigated the maintenance of supersonic combustion with the aid of an attached microwave discharge in a plane flow.

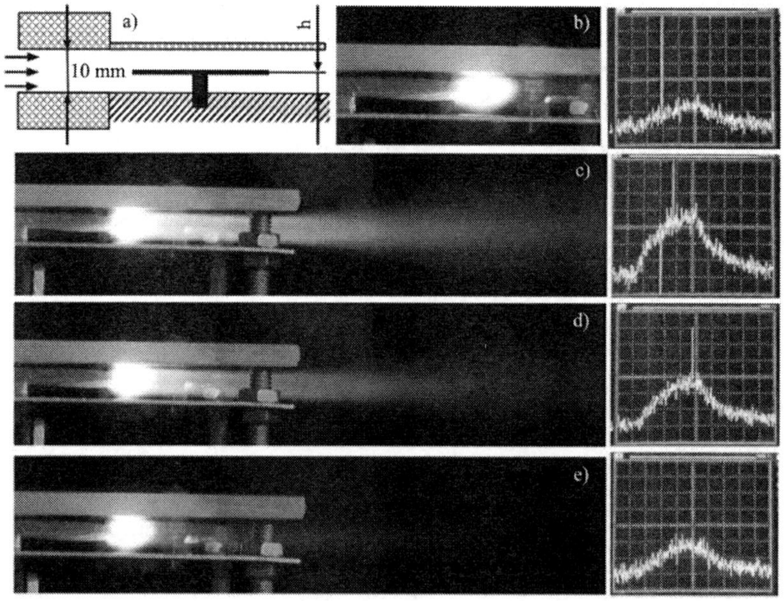

Fig. 4. The connected discharge in the mixture of propane with air at a flow rate of 30 m / s: a) – the discharge initiator scheme, b) - discharge in the air, c) combustion of the stoichiometric mixture, fuel surplus factor $\alpha = 1$, initial pressure p = 3.3 bar; d) - combustion of the lean mixture, $\alpha = 0.7$, p = 1 bar; e) - combustion of a very poor mixture at $\alpha = 0.5$, p = 0.3 bar. h = 5 mm. Vertical division scale – 100 °C.

The results of the experiment for propane injection into the wake air flow through the tubular resonator are shown in Fig. 5. It is seen that, despite the weakening of the flame, combustion continues at supersonic speeds.

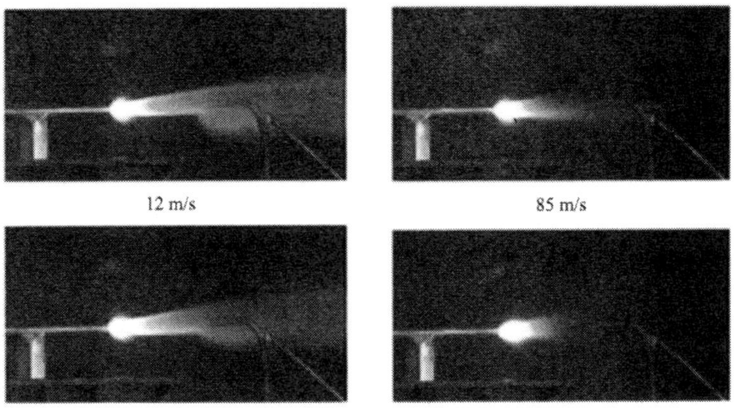

12 m/s 85 m/s

30 m/s 200-500 m/s

Fig. 5. Combustion of propane in the cocurrent air flow by means of a subcritical microwave discharge at various airflow rates.

In Fig. 6 shows the temperature field in the cross section of the flow at a distance of 1.4 cm from the feed cut of the mixer. Data are taken for supersonic flow velocity. Noticeable temperature increase of 70-100 degrees Celsius.

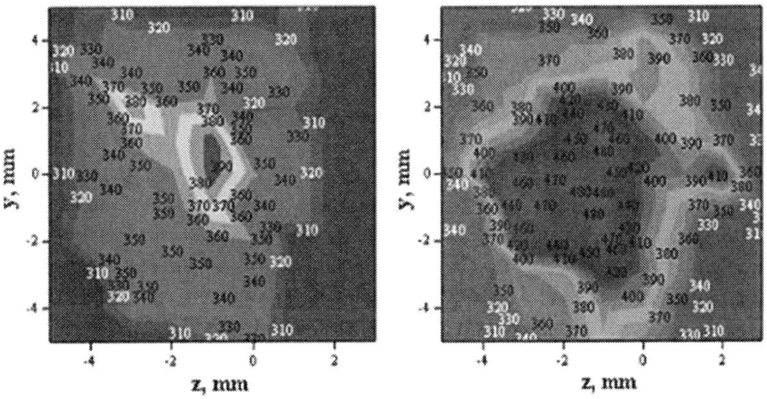

Fig. 6. Temperature in the cross section of the supersonic flow at a distance of 1.4 cm from the feed cut of the mixer; (a) - air is injected into the discharge area through the mixer, (b) - an air-propane mixture is injected into the discharge area.

6 Conclusion

Calculations have shown that the velocity of the streamers is several km / s, so they are not blown off by a supersonic flow. Experiments have been conducted to ignite the flow of a mixture of propane with air by various discharges at speeds up to twice the speed of sound. In all cases, the experiments confirmed a steady burning of the fuel, which was controlled from the temperature measurements with a thermocouple.

The work was supported by the Ministry of Education and Science of the Russian Federation (agreement No. 14.577.21.0277, unique identifier of the project RFMEFI57717X0277).

References

1. P.V. Bulat, I.I. Esakov, L.P. Grachev, P.V. Denissenko, M.P. Bulat, I.A. Volobuev, Sc. and Tech. J. of ITMO **17**, 569 (2017)
2. M.P. Bulat, P.V. Bulat, P.V. Denissenko, I.I. Esakov, L.P. Grachev, K.N. Volkov, I.A. Volobuev, Acta Astronaut. (2017)
3. A.M. Hovatson, *An Introduction to Gas Discharges* (Pergamon Press, 1976)
4. K.V.Khodataev, B.R.Gorelik, Plasma Phys. **23**, 236 (1997)
5. V.V. Upyrev, I.I. Esakov, L.P. Grachev, Noneq. Proc. in phys. and chem., **1**, 208 (2016)

Increase of the energy plant efficiency in special conditions of its operation

Razuvaev A.V.[1], *Biryuk V.V.*[2], and *Blagin E.V.*[2*]

[1]Balakovo Engineering-Technological Institute, 413853 Balakovo, Russian Federation
[2]Samara University, 443086 Samara, Russian Federation

Abstract. This article deals with energy plant analysis during its operation in special conditions. For this purpose, modernized system of air supply and high-temperature refrigeration is suggested. Application of these systems allows to decrease specific fuel rate by 2.6%.

1 Introduction

Development of the arrangements for cost reduction of the internal combustion engines (ICE) power plant exploitation for consumer in market conditions of Russian economics is an actual task. This fact forces ICE manufacturers to decrease cost of available products during its realization and production for expansion of the manufacturing and selling area.

Increase of the exploitation economic efficiency and durability with taking into account special condition of ICE exploitation is an actual task (besides improvement of the manufactured goods quality) which provide competitive ability in comparison with native and foreign similar products.

These tasks can be solved based on achievement of certain parameters, which determine economic efficiency of certain models of ICE. This article deals with solution of such tasks based on experience of ICE experimental research at manufacturers' facilities and during their exploitation.

Energy plants of different purpose on the base of piston ICE operate on different regimes starting from idle run and up to nominal power output, and in case of an engine used as a generator drive, ICE may operate on maximal regime, which is 110% of nominal power output [1].

For example, utilization of the energy plant as the energy source for diesel locomotive used primarily for road switching services on nominal regime requires 0.5-1% of power capacity and on partial loads and idle run – up to 60% [1]. For locomotives used in regional rails, this value reaches 3-10% on nominal regime, up to 30% on the regime of partial load and average power output is 70-80%.

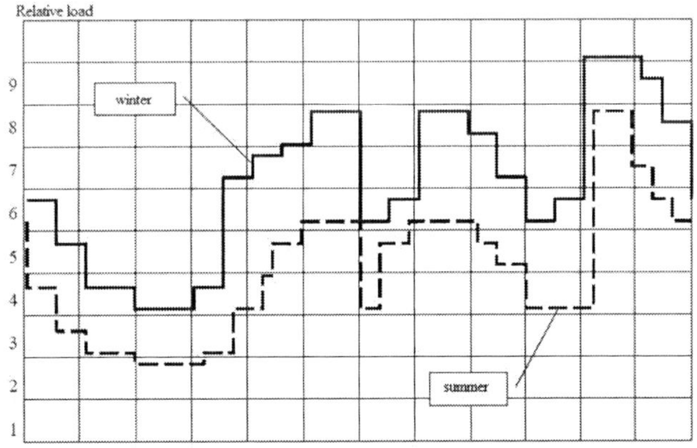

Fig.1. Plot of the day load of diesel energy plant

Figure 1 shows a plot of energy consumption by a neighbourhood [2] with developed industry and culture (theatres, cinema, etc). Usual minimum of energy consumption is located between 4 and 6 o'clock, maximum is located between 8 and 16-17o'clock with a gap from 11-12 o'clock and 13-14 o'clock with peak consumption between 19-20 o'clock and 23-24 o'clock. In different time of year, energy, water and gas consumption also varies and reaches maximum at winter

* Corresponding author: evgenyblagin@gmail.com

and minimum at summer (for gas and electricity) and vice versa for water. Such periodicity is inevitable; it is utilized as a factor that allows maintaining technical condition of the energy plant on necessary level.

Exploitation of the "Slavyansk" and "Bezhitsa" ships with tree diesel generators shows that there is a possibility to use one diesel generator on the march if conditioner is not working. In case of standing in tropical climate during cargo operations, load of the energy plant was 220-230 kW, during operation load increased up to 260-270 kW. Thus, if two diesel generators were working, load of each one was no less than 50% of nominal power.

Due to this, if one diesel generator with load of 270 kW is operating instead of two operating in parallel regime, fuel economy per day can reach 0.8 tons and oil economy can reach 8-10 kg. Engine's overhauled period remains, maintenance cost is decreased. Peaks of short time load up to 300 kW during operation on standing in tropical climate does not allow to utilize just one diesel generator because of automatic overload control is usually set for 260-270 kW.

Thus, for optimization of the energy plant load, it is recommended to apply automatic overload protection, set on 275-280 kW with 8-10 minutes exposition and at 290-300 kW it is necessary to turn off secondary consumers with 5-10 seconds exposition.

Operation of energy plant of BelAZ-540 truck with D12-375E diesel generator nominal power is utilized during 5...8 minutes.

Besides, exploitation of energy plants of different purpose is performed in different climate conditions. In conditions of increased ambient temperature ICE power must be decreased according to exploitation manual.

There are also objects which has an increased resistance on suction and exhaustion in comparison with normal exploitation conditions. In this case, decrease of the maximal or nominal power of ICE is also required. Here, it should be noted that short time in overload regime requires providing of necessary power by means of power increase and this cause increased cost of the energy plant and entire facility.

Utilization of the energy plant on transport does not always allow to increase its power because it causes increase of the dimensions and weight of the vehicle.

That's why to solve these tasks and provide relatively short overload periods in relation to nominal or maximal power without decrease of the ICE durability, modernized system of air supply (MSAS) is suggested. This system considers steam injection in turbo-compressor which provides increase of its power and also increase air supply in ICE cylinders. Turbo-compressor power is increased by working fluid flow rate increase in the form of steam-gas mixture. Temperature of the steam-gas mixture is decreased because steam has lower temperature than exhaust gas. Further, increase of the air flowrate through ICE causes decrease of the specific fuel consumption and temperature of the exhaust gases, which in turn increases durability of entire power plant.

System itself consists of simple equipment such as water storage tank, which can be located in any suitable place and have almost any configuration and its volume is estimated from necessary water margin considering real operation regimes of exact energy plants. Water pumps, which provide water pumping from water margin tank into steam generator. Generated steam is pumped through the valve by duct on selected area of the turbocompressor nozzle vanes. Heat source for steam generation is a heat of the exhaust gases. Dimensions of the steam generators are determined from necessary steam flow rate for its charge in ICE.

Having such system gives an opportunity to inject small portion of steam on transient regime (i.e. during power increase) which provide transient process duration decrease and also pollution exhaust decrease from incompleteness of combustion. This fact causes increase of the energy plant ecological characteristics.

Further, analysis [4] shows that additional effect can be expected from MSAS system application. During exploitation COP of turbo compressor is decreased due to decrease of the gas turbine COP due to accumulation in its flow area of carbon and combustion incompleteness products. Observations shows that carbon layer thickness is stabilized after 400.500 hours of engine exploitation. That's why for this period charging pressure is decreased on 8-10%, compressor COP – on 15...20% which causes increase of the gas temperature on 10...15% and specific fuel rate on 3...6% (6...11 g/kW h).

To keep gas-turbine boost in good technical condition, "Brown, Boveri", "Ishikawajima-harima" machinery companies apply cleaning of the gas-turbine boost on operating diesel [4]. This method of gas-turbine boost cleaning allow to operate without disassembling up to 12...16 thousand hours instead of 5...6 thousand hours according to manufacturer's recommendation and coincide disassembling with bearing replacement.

Methods of cleaning developed by "Brown, Boveri" and "Burmeister & Wain" companies have a most wide distribution.

Method of "Brown, Boveri" company is that gas turbine cleaning is executed on low frequency of the diesel. Temperature of the cleaning water is kept within 323...333 K, cleaning periodicity is 200.250 hours and cleaning duration is 10...20 hours and depends on carbon accumulation degree. Cleaning is performed by clean fresh water.

Method suggested by "Burmeister & Wain" company is that turbine and compressor cleaning is peformed by fresh water with 0.2...0.5 MPa pressure which is sprayed by air with 0.4...1.0 MPa pressure. Cleaning of the gas part and the compressor is performed under decreased rotation frequency. Cycle of cleaning for turbine is 20...25 minutes after 100.150 hours and for compressor is 5 minutes after 24...48 hours of its operation.

In Far-Eastern Shipping Company both methods of cleaning are applied with success on types "Pioneer", "Mikhail Kalinin" and "Vyborg" ships since 1970.

On the base of aforementioned we can suggest that application of MSAS similarly to cleaning of the diesel turbocompressor flow part will additionally promote increase of the diesel application efficiency in term of maintaining the given value of turbocompressor COP and increase diesel operation duration between repair.

Another way to increase the power plant efficiency is high-temperature refrigeration (HTR). Application of HTR during ICE exploitation gives certain positive effect, which is decrease of the mass-dimensional characteristics to 8-20%, which is important for power plants of transport vehicles and for their placement as a drive for different equipment. HTR system equipment is not complicated but this regime promotes increase of temperature of the sleeve assembly elements. This fact can slow down application of HTR because every degree increment of the refrigeration liquid temperature causes the increase of the sleeve assembly temperature on 0.8 degree. That's why application of arrangement which provide decrease of the sleeve assembly in HTR regime is an important task.

2 Calculation algorithm

Analysis applied above as well as founded advantages of the MSAS system testifies the reasonability of knowledge about ICE with MSAS parameters. For this reason, calculation method is suggested. Purpose of calculation is determination of the steam amount, which can be obtained from the exhaust gases heat and charged to flow area of the gas turbine of turbocompressor as well as determination of ICE parameters changed due to MSAS application. Let's estimate an amount of steam generated by heat utilization from exhaust gases considering heat utilization from refrigeration water:

$$G_S = \frac{G_g c_{pg} T_{g2} - G_g c_{pg} T_{g3}}{r + c\left(T_s - T_{rw}\right)} \tag{1}$$

where $T_{g3} > 423$ K (selected from requirement for inhibition of gas duct acid corrosion);

T_{rw} – reserve water temperature of the water storage is determined according to water temperature on the exit from ICE.

Then we select steam pressure and temperature correspondingly with condition $p_s = p_{g1} + (0.02...0.03)$ MPa. Steam flow rate value obtained from expression (1) is checked taking into account average gas pressure before the turbine pg1, steam pressure and temperature (ps and Ts) and duct geometric dimensions (l, d, Fs):

$$G_s = F_s \sqrt{\frac{p_s^2 - p_{g1}^2}{(\lambda \frac{l}{d} + 2\ln\frac{p_s}{p_{g1}}) R_s T_s}} \tag{2}$$

Diesel parameters calculation during its operation with MSAS starts with determination of the steam gas mixture flowrate through the turbine

$$G_{sg} = \left(G_{air} + B\right) + G_s. \tag{3}$$

For the first step air flow calculation through the diesel during its operation with MSAS and HTR is the same: $G_{air} = G_{air}^{sg}$.

Weight parts of the steam gas mixture components are found as:

$$g_s = G_s / G_{sg} \qquad g_g = G_g / G_{sg}. \tag{4}$$

After this, heat capacities and gas constant of steam gas mixture are determined:

$$c_p^{sg} = c_p^g g_g + c_p^s g_s \qquad R_{sg} = R_g g_g + R_s g_s \tag{5}$$

Temperature of the steam gas mixture on the entrance to the turbine of the turbocompressor is calculated as:

$$T_{g1}^{sg} = \frac{g_s c_p^s T_s + g_g c_p^g T_{g1}}{g_s c_p^s + g_g c_p^g} \tag{6}$$

Pressure drop of the working fluid on the turbine is proportional to its velocity and specific weight:

$$\Delta p \sim \frac{\omega^2}{2g}\gamma \sim \frac{V^2}{2g}\gamma \sim \left(\frac{G^2 R^2 T^2}{p^2}\right)\frac{p}{RT} \sim \frac{G^2 RT}{p} \tag{7}$$

Considering this, if Δp_T of the turbine operation on exhaust gases is known, for the case of MSAS:

$$\Delta p_t^{sg} = \Delta p_t \frac{(G_{sg})^2 R_{sg} T_{g1}^{sg} p_{g1}}{(G_g)^2 R_g T_{g1} p_{g1}^{sg}} \tag{8}$$

In this expression for the first intermediate calculation step we accept that $p_{g1} = p_g^{sg}$. Value of Δp_t^{sg}., founded this way, is utilized for elaboration of the value:

$$p_{g1}^{sg} = \Delta p_t^{sg} + \Delta p_{exh} \tag{9}$$

Calculation by expressions (8) and (9) is executed by step by step method and is stopped when deviation in accepted and calculated values of p_{g1}^{sg} will be less than 2 kPa. Further, next values are determined:

$$\Pi_T = \frac{p_{g1}}{p_{g2}} \text{ and } \Pi_T^{sg} = \frac{p_{g1}^{sg}}{p_{g2}^{sg}} \tag{10}$$

For engineering calculation it can be accepted that (it is verified by experiment) transition to MSAS is followed by almost similar increment of the expansion ratio in turbine and compression ratio in compressor. It allows to determine values of Π_c^{sg} and p_{int}^{sg}:

$$\Pi_c^{sg} = \Pi_c \frac{\Pi_T^{sg}}{\Pi_T} \qquad (11)$$

$$p_{\text{int}}^{sg} = \Pi_c^{sg} p_d. \qquad (12)$$

Air flowrate change through the diesel will be proportional to change of the charge air pressure because air temperature in induction manifold is almost constant:

$$G_{air}^{sg} = G_{air} \frac{p_{\text{int}}^{sg}}{p_{\text{int}}}. \qquad (13)$$

As it known, heat equivalent of turbine adiabatic work is:

$$Q_{adT}^{sg} = \frac{k_{sg}}{k_{sg}-1} R_{sg} T_{g1}^{sg} G_{sg} \left[1 - \left(\frac{1}{\Pi_T^{sg}}\right)^{\frac{k_{sg}}{k_{sg}-1}}\right]. \qquad (14)$$

From the other side, heat content and the exhaust gases temperature before the turbine at MSAS will be equal to:

$$Q_{g1}^{MSAS} = Q_{g1} + \left(G_{air}^{sg} - G_{air}\right) c_{pair}.T_d; \qquad (15)$$

$$T_{g1}^{MSAS} = \frac{Q_{g1}^{MSAS}}{\left(G_{air}^{sg} + B\right) c_{pair}}. \qquad (16)$$

Heat content of the steam gas mixture before turbine can be found as:

$$Q_{g1}^{sg} = Q_{g1}^{MSAS} + G_s c_{ps} T_s. \qquad (17)$$

To check, this value can also be found as:

$$Q_{g1}^{sg} = G_{sg} c_{psg} T_{g1}^{sg}. \qquad (18)$$

Then heat content and temperature of the steam gas flow after the turbine is calculated as:

$$Q_{g2}^{sg} = Q_{g1}^{sg} - Q_{agT}^{sg}; \qquad (19)$$

$$\mathrm{T}_{g2}^{sg} = \frac{Q_{g2}^{sg}}{G_{sg} c_{psg}}. \qquad (20)$$

Accounting founded values of $G_{air}^{sg}, T_{g2}^{sg}, G_s$, diesel parameters calculations starting from expression (1) are repeated and to determine the value of G_s, values of $G_{sg}, c_{psg}, T_{g2}^{sg}$ are substituted instead of values of G_g, c_{pg}, T_{g2}. Difference between the air flowrate and gas temperature between current and previous calculation must be within the range of ±(2...3) kg/h and ±(4...5)°C.

Value of α_{sg} on the first step of the calculation is determined as:

$$\alpha_{sg} = \frac{G_{air}^{sg}}{B \cdot 14.35} \qquad (21)$$

assuming constant flowrate and equal to fuel rate at diesel operation with HTR.

Diesel crank shaft frequency and average oil temperature in the lubricating system do not change after transition to MSAS, so mechanical COP can be considered constant. Thus, using known by manufacturing tests dependence $\eta_i = f(\varepsilon)$ value of η_i^{sg} is determined according by the value of α_{sg} founded above. Then we found the value of specific fuel rate and corresponding hour fuel rate B_{sg}. Considering founded value of B_{sg} by expression (21) values of α_{sg}, η_i^{sg} and η_e^{sg} are specified.

Finally, specific fuel rate at MSAS is found as:

$$B_{sg} = B \frac{\eta_e}{\eta_e^{sg}}. \qquad (22)$$

Example of calculation for the diesel "6ChN21/21" with MSAS is presented in table 1.

Initial data for the diesel calculation with MSAS are experimental values of diesel operation with HTR with load of 525, 627 and 845 kW by load characteristics with crank shaft frequency of n=25 Hz.

3 Calculation results

Analysis of the data obtained at 845 kW power shows that with the calculated steam flow rate value of 0.196 kg/s (10.5%), charged in the turbine of turbocompressor, decrease of the specific fuel rate is equal to 6 g/kW h (2.6%), hour fuel rate decrease is 5 kg/h, working fluid temperature before turbocompressor decrease is 87 oC (9.3%), increase of the air flow rate is 0.092 kg/s (6%) and increase of the air-fuel ratio – 0.17 (8.6%).

Table 1. Calculation results.

Parameter	Units	P_e=525 kW		P_e=627 kW		P_e=845 kW	
		HTR exp.	MSAS calc.	HTR exp.	MSAS calc.	HTR exp.	MSAS calc.
Steam flow rate	kg/s	-	0.123	-	0.1685	-	0.196
Exhaust gases flow rate	kg/s	1.145	1.19	1.35	1.41	1.578	1.671
Steam gas mixture flow rate	kg/s	-	1.31	-	1.578	-	1.867
Mass part of the steam in steam-gas mixture	%	-	9.4	-	10.7	-	10.5
Mass part of the gas in steam-gas mixture	%	100	90.6	100	89.3	100	89.5
Temperature before turbo compressor	K	824	760	883	803	933	846
Pressure before turbo compressor	MPa	0.19	0.199	0.223	0.234	0.278	0.298
Expansion ratio		1.82	1.895	2.065	2.173	2.574	2.766
Compression ratio in comrpessor		1.9	1.978	2.3	2.42	2.989	3.21
Air pressure in induct manifold	MPa	0.19	0.1978	0.228	0.239	0.293	0.321
Air flow rate	kg/s	1.11	1.15	1.307	1.373	1.525	1.617
Temperature after the turbine	K	731	642	759	667	788	658
Total air-fuel ratio		2.18	2.33	2.08	2.23	1.96	2.13
Indicated efficiency	%	46.8	47.8	46	47.2	45.1	46.4
Specific effective fuel rate	g/kW*h	240.4	235.3	233.9	227.9	230.5	224.5

4 Conclusion

Analysis of the possible regimes of energy plant operating is performed on the base of piston ICE in special conditions, which differ from the conditions, specified in typical functional specifications. Opportunity of MSAS application is suggested, which allows to increase or at least maintain energy plant durability during its operation in special conditions. Calculation method is suggested which allows to determine parameters of the ICE with MSAS during its operation in special condition on the design stage. It also allows getting initial data for design and technological works by creation of MSAS system equipment. Calculation result of ICE parameters during its operation with modernized system of air supply shows that its economic efficiency on given level of power output is increased which causes increase of the exploitation economic efficiency of power plant and simultaneously decrease of hazardous working fluids exhaust together with exhaust gases from not burned fuel. Decrease of the steam-gas mixture and increase of the air flowrate through ICE promotes decrease of the temperature of sleeve assembly elements which promotes increase of the durability of power plant in special conditions. MSAS system application on ICE during its operation in special condition is justified and further development of this problem is relevant in the field of calculation of the necessary equipment parameters and algorithm determination of the design-technological parameters of the system elements.

References

1. A.V. Razuvaev, Piston internal combustion engines with high-temperature refrigeration (Saratov state technical university, Saratov, 2001) (in Russian)
2. F.L. Liventsev, Energy plants with internal combustion engines (Mashinostroenie, Leningrad, 1969) (In Russian)
3. M.K. Ovsyannikov, V.A. Petukhov, Fuel utilization efficiency in ship diesel plants (Shipping, Leningrad, 1984) (In Russian)
4. B.I. Soloviev, V.D. Ivanov, P.S. Emelianov. MAN diesel exploitation (Transport, Moscow, 1978) (in Russian)
5. A.V. Razuvaev, Turbines and Diesels (in Russian) *3(30)* (2010)

The advisability of using combustion chambers with a toroidal recirculation-mixing zone in small-sized gas turbine engines

Michael Yurievich Orlov[1], *Oleg* Vladimirovich Kolomzarov[1], *Vladislav* Mikhailovich Anisimov[1 1,], *Nikita* Igorevich Gurakov[1] and *Nikolai* Sergeevich Mironov[1]

[1]Samara National Research University, Heat engine Department, Samara, 443086, Russia

Abstarct. The interrelations between the size of the gas turbine engine (GTE) and the size and workflow of the combustion chamber (CC) were considered. On the base of analysis of workflow organization the design of CC with toroidal recirculation mixing zone was proposed. The theoretical justification of chosen design was carried out on the base of comparison of combustion volumes, which formed in traditional CC with swirlers and in proposed CC. The comparison of combustion volumes, schematic display of combustion zones with a discrete flame and a combined combustion zone were given.

A large number of different requirements can be offered to small-sized GTE (weight, size, durability, level of environmental characteristics, etc.) due to the wide range of tasks it solve. The fulfillment of most of these requirements is related to the choice of the combustion chamber design and is one of the most important problems of preliminary design.

The feature of small-sized engines is its more strict interrelationships between workflow parameters and design. As a result, in such cases, a slight change in geometry can cause a flow reversal. Therefore, with a reduction of the size of the engine relative to the prototype, its combustion chamber cannot be reduced by simply scaling its elements, because of the changes in the preparation of fuel-air mixture, the organization of the cooling system and air mixing.

When considering the combustion chambers of small-sized engines, it is important to determine the meaning of the term "small size", because of different approaches.

The analysis of these criterions shows that, in common case, the engine with air flow rate within the range $1 \leq G_f \leq 10$ kg/s can be referred to the small-sized engines. For impeller machines of small-sized GTE is characterized by the decrease of power efficiency, and its combustion chamber is characterized by a significant reduction in geometric size and workflow parameters (inlet pressure, outlet temperature, combustion efficiency). The characteristic feature of these engines is the presence of the last centrifugal or axial-centrifugal compressor stage and, as a consequence, the availability of various combustion chamber designs.

As parameters characterizing the dimension, the following applies:

[1] Corresponding author: vradik@mail.ru

- air flow rate reduced by parameters beyond the compressor [1]:

$$G_{\text{f red}} = G_{\text{f}} \frac{101325}{p_{\text{c}}^{*}} \sqrt{\frac{T_{\text{c}}^{*}}{288}}; \tag{1}$$

- air flow rate complex:

$$G_{\text{cc}} = \frac{G_{\text{f}} \sqrt{T_{\text{c}}^{*}}}{p_{\text{c}}^{*}}; \tag{2}$$

- air mass flow rate;
- the air flow rate reduced to the parameters behind the compressor:

$$G_{\text{f 1red c}} = \frac{G_{\text{f1 0}}}{\pi_{\text{c}\Sigma}^{*\,5/6}}. \tag{3}$$

There is another approach to the determination of concept of small-size known from theory of GTE [3]. The specific weight was considered in [3] as a function of the characteristic size of the engine (D):

$$\gamma_{\text{engine}} = \frac{M_{\text{engine}}}{N} \approx f\left(\frac{D^3}{D^2}\right) = f(D), \tag{4}$$

where γ_{engine} – engine specific weight, M_{engine} – engine weight, N – engine power, D – linear engine size.

In paper [4] graphical representation of the dependence (4) is shown, where we can see that theoretical graph coincides with statistical data only in a certain range (fig. 1). This is due to the fact that as the size of the GTE decreases, not all nodes and systems manage to scale proportionally.

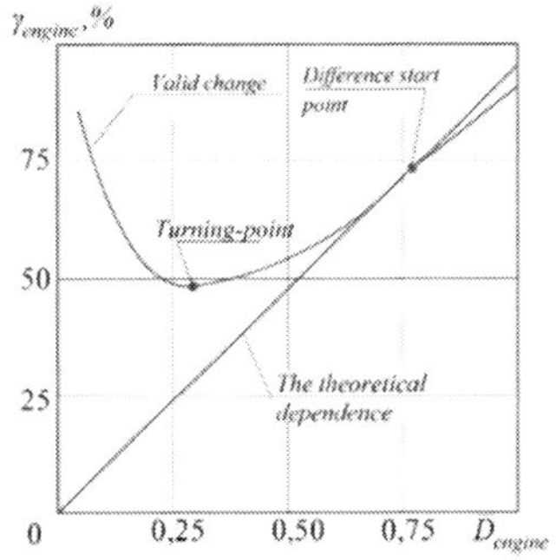

Fig 1. Dependence of specific weight (γ_{engine}) on the relative diameter of the engine.

Hence, starting with some dimensions of the CC, when they are further reduced, the proportionality in the interrelation between the parameter of the cycle of the gas turbine engine (π_{c}) and the optimal conditions for the workflow in the CC (τ_{ts}) is violated.

As the main functional and technological reasons that limit the dimensions of combustion chambers small-sized GTE can act [4,5]:

- technological tolerances for the dimensions of all elements of the CC;
- the sizes of channels for fuel supply;
- dimensions of mixing elements – burners (nozzles)
- gaps between moving elements;
- diameter of fuel droplets;
- the thickness of the gas curtain near the walls of the flame tube;

- non-uniformity of distribution of parameters and substances along the cross-section of the flame tube.

On the other hand, a number of parameters do not change their size with decreasing dimensions of the CC:

$$\tau_{burn} \geq \tau_{cr} + \tau_{evap} + \tau_{mix} , \qquad (5)$$

where τ_{burn}, τ_{cr}, τ_{evap}, τ_{mix} – times of burning, chemical reaction, evaporation and mixing, respectively.

The geometric limitations imposed on the choice of the scheme and design of the CC are also dictated by the ratio of the diametrical dimensions of the last stage of the compressor, the first stage of the turbine and the axial dimension between them. These parameters determine the length of the combustion chamber, the angle of its tilt, the overall size of the body and other geometric characteristics. Since one of the basic requirements for the organization of the workflow of the CC is to create conditions for flame stabilization and ensure high completeness of combustion in the primary zone, it can be argued that the limitations on the dimensions imposed by the compressor and turbine on the geometry of the CC are ultimately restrictions on the organization of its workflow.

For the sustainable combustion in each design of CC requires a certain set of structural elements (nozzles, swirlers, etc.) with a certain mutual arrangement and a combination of sizes. At the same time, each of the constructive ways of organizing combustion has its own peculiarities and, accordingly, the possibilities of using it in each specific case for implementing the corresponding scheme of the workflow.

At present, various designs of the flame tube are used to distribute fuel through the primary air layers: with a swirler, a hemispherical front device of the louver type, and several others.

As an alternative to these methods of organizing combustion, a method of forming a toroidal recirculation zone can be used, with a rotating vortex capable of stabilizing the focus of the flame [4]. This method allows us to: use cheaper and easier to manufacture jet nozzles as the main ones, reduce the height of the flame tube and increase the compactness of the compressor.

The authors in [4] provides a theoretical computation of the justification for this method of organizing combustion. It reduces to determining the criterion for stable combustion in a CC by means of the ratio of the energy required for heating the mixture E_r and the energy available for this process E_a.

$$E_e = C_1 \cdot (D_{burn})^3, \qquad (6)$$

$$E_r = C_2 \cdot (D_{burn})^2. \qquad (7)$$

From this we obtain the relation

$$\bar{E}_r = \frac{E_r}{E_e} = \frac{C_3}{D_{burn}} . \qquad (8)$$

From which it can be seen that minimization of the autonomous combustion source in the CC has a limit in the form $\bar{E}_n = 1$.

When comparing the volume of the reverse flow zone of an individual burner and combustion chamber with a toroidal recirculation zone, if we assume that an individual burner (vortex or other type) creates a cylindrical combustion zone with dimensions $D_{bf} = C_4 h_{ft}$ (h_{ft} – height flame tube) and l_{bf}. Then

$$\left(V_{\mathrm{bf}}\right)_{cil} = C_5 \cdot l_{\mathrm{bf}} \cdot \frac{\pi \cdot \left(h_{\mathrm{ft}}\right)^2}{4} \tag{9}$$

Then, if we know the coefficient C_5, we will be able to determine the maximum value of the height of the flame tube h_{ft}, to which it is possible to reduce the dimensions of the CC without risk of reaching the limit (7). This value is the boundary in size of the burner - $(D_{\mathrm{burn}})_{\mathrm{min}}$ and approximately equal to 15-20 mm [4,6].

To go beyond this limitation, you can combine the discrete combustion zones with the volume $(V_{bz})_{cyl}$, located along the ring around the axis of the engine, into a single annular combustion zone (fig. 2). If we assume that the combustion zone of such a "burner" has the form of a torus, then its size can be written as:

$$\left(V_{\mathrm{bz}}\right)_{\mathrm{thor}} = C_6 \cdot 2\pi R \cdot \frac{\pi \cdot \left(h_{\mathrm{ft}}\right)^2}{4} \tag{10}$$

where R – large radius of the torus, and h_{ft} is the height of the flame tube.

However, the limits of flame stabilization can be considered in other ways. Thermal theory:

$$\frac{q_2}{q_1} = \frac{w^{n} d^{n} (T_{\mathrm{mg}} - T_0)}{\nu^{n} w (T_{\mathrm{f}} - T_0)} = const, \tag{11}$$

where $q1$ – amount of heat transmitted by the fresh mixture from the combustion products of the circulation zone, $q2$ – amount of heat required by the fresh mixture for ignition. Contact theory

$$\frac{t_1}{t_2} \sim \frac{d \cdot U_{\mathrm{H}}^2}{w \cdot a_{\mathrm{m}}} = \mathrm{Mi}. \tag{10}$$

This theory is based on the ratio of the times of contact of the fresh mixture with the hot combustion products in the recirculation zone t_1 and the time necessary for heating the mixture to the temperature required for ignition. Reactor theory is also existed [5].

On the basis of these theories, we can assume that stabilization mainly depends on the speed of air and fuel-air mixture, the thermal diffusivity, the ignition temperature and the temperature of the mixture, and also on the size of the recirculation mixing zone. If all other parameters are assumed to be constant, a zone with a larger size is obtained with greater stability, which makes it possible to provide either a reserve of available energy or to form an area in which there will be a longer contact time.

If we consider the flow in a scheme with a swirler and a toroidal one, then if the holes of the first row are identical, it is evident that in the second case preconditions exist for the organization of a more developed combustion zone (fig. 2).

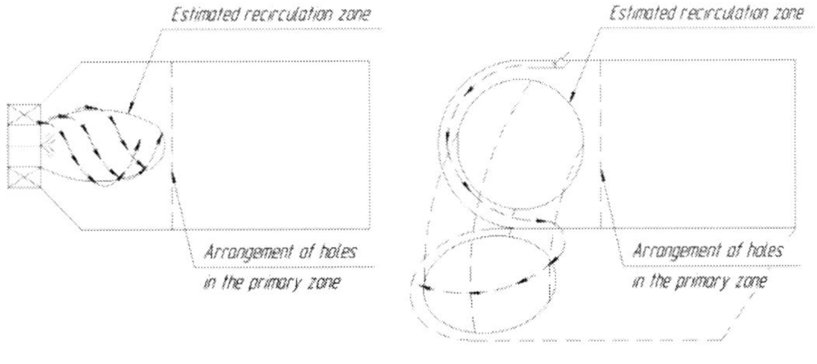

Fig. 2. Comparison of particle tracks of different recirculation zones.

Thus, the toroidal form of the combustion zone has the advantage from the point of ensuring stable combustion. However, its application raises a number of new tasks related to the dynamics of gas flow, the search for optimal ways of preparing fuel assemblies and with questions of energy-mass transfer. Therefore, for the further investigation, numerical modeling of the working process was used.

On the basis of the above theoretical calculations, within the implementation of one of the projects, a combustion chamber was developed, the initial shape of which is shown in fig. 3.

Further, the investigation of the CC workflow will be conducted by simulation method to prove the toroidal scheme of combustion.

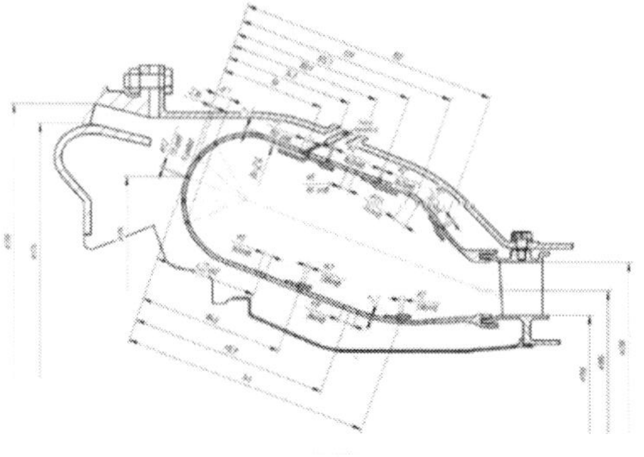

Fig. 3. The general view of CC.

Discussion

The toroidal recirculation zone, with the correct organization of combustion, can contribute to a more uniform exit of hot gas from the back-current zone, due to the absence of a discrete flame source, and accordingly to a more uniform temperature field in the circumferential direction at the outlet from the CC.

The comparison of the dependences of the volume of combustion zones for two types of combustion organization showed that the toroidal recirculation zone makes it possible to reduce the diametrical dimensions of the flame tube with the same V_{bz}. This flame tube design allows forming the toroidal recirculation-mixing that leads to increasing the path of burning mixture.

In addition, the lack of the traditional flame tube head will reduce the cost of production of small-sized GTD.

This work was supported by the Ministry of education and science of the Russian Federation in the framework of the implementation of the Program "Research and development on priority directions of scientific-technological complex of Russia for 2014-2020" (RFMEFI58716X0033).

References

1. V.A. Grigorev, V.V. Zrelov, Yu.M. Ignatkin, *Vertoletnie gazoturbinnie dvigateli*, 490 (2007).
2. V.S. Kuz'michev, Vestnik Samarskogo universiteta. Aehrokosmicheskaya tekhnika, tekhnologii i mashinostroenie, **2**, 15 (2016).
3. A. Dzhadzh, *Gazoturbinnye dvigateli maloj moshchnosti*, 270 (1963).
4. V.P. Lukachyov, Vestnik Samarskogo universiteta. Aehrokosmicheskaya tekhnika, tekhnologii i mashinostroenie, **1**, 1-15 (1998).
5. A.M. Lanskij, Metody i sredstva povysheniya ehffektivnosti rabochego processa kamer sgoraniya malorazmernyh GTD: diss. (2014)
6. A.M. Lanskij, S.V. Lukachev, S.G. Matveev, SNC RAN, **2**, 1-11 (2009)

Effect of secondary swirl in supersonic gas and plasma flows in the self-vacuuming vortex tube

Vyacheslav Volov[1]1, *Anton* Lyaskin[2]

[1]Department of Natural sciences, Samara State Transport University, Samara, Russia
[2]Department of Computational Fluid Dynamics, CADFEM-CIS, Samara, Russia

Abstract. This article presents the results of simulation for a special type of vortex tubes – self-vacuuming vortex tube (SVT), for which extreme values of temperature separation and pressure drop are realized. The main results of this study are the flow structure in the SVT and energy loss estimations on oblique shock waves, gas friction, instant expansion and organization of vortex bundles in SVT. Keywords: self-vacuuming vortex tube, supersonic swirling gas flows, oblique shock wave, rotating cord.

Nomenclature

t – time, x_i – Cartesian coordinates, u_i – Reynolds-averaged components of instantaneous velocity vector in the Cartesian coordinate system, u_i' – pulsation components of instantaneous velocity vector in the Cartesian coordinate system, ρ – density, p – pressure, h_t – total enthalpy, T – temperature, λ – coefficient of thermal conductivity, c_p – specific heat capacity at constant pressure, μ – dynamic viscosity, $\rho\overline{u_i'u_j'}$ – components of turbulent stress tensor, k – specific turbulent kinetic energy, ω – specific rate of dissipation of k, P^* - total (stagnation) pressure, T^* - total (stagnation) temperature.

1. Introduction

Vortex flows have been of significant interest since the mid-20th century because of their occurrence in industrial applications (Gupta et al. [1]). For examples, different vortex and swirling devices can be used in gas producing units of gas turbine engines ([2], [3]). Besides, vortex (or high degree of swirling) can also produce a hot and cold stream separation via Ranque-Hilsch vortex tube [4, 5]. Nash [6] and Dobratz [7] provided extensive reviews of vortex tube utilizations and enhancements. Intense experimental and numerical studies of Ranque-Hilsch tubes have begun since 1931 and continue even nowadays [8, 9, 10]. The most famous scientific school studying the Ranque-Hilsch effect in Russia was led by A.P. Merkulov [11, 12, 13]. In addition to traditional applications of vortex tubes, this school proposed for them a number of new utilizations in aviation, chemical industry, power industry, agriculture, etc. Special merit of this scientific school is the creation of vortex electro-discharged devices: CO_2-lasers and plasmatrons [14, 15]. A significant number of theories have been proposed to explain the vortex effect of «temperature separation» since its initial observations by Ranque [4]. Despite all proposed theories, none has been able to fully explain the one. This fact is partially compensated by the numerical models of flow in vortex tubes. Numerical studies of swirling gas flows in the vortex tube were presented by following researches [8, 9, 10]. Simulations were made using turbulent models ranging from simple RANS (as standard k-ε) [8] to large eddy simulation (LES) [9]. At the same time the least developed type of vortex tubes in theoretical and experimental aspects is the self-vacuuming vortex tube (SVT), where extreme values of temperature separation (ΔT_{cold}=152K, gas - air [1]) and vacuum ($\pi^*=P^*/P_{axis}$=40, where P^*- total gas pressure at the inlet of SVT, P_{axis}- gas pressure near the axis of SVT) are be realized. In contrast with classic vortex tubes [4, 5] SVT do not have hot and cold flow exits. Instead, SVT is used for cooling of its internal body. There are some engineering models for calculation of flows and thermodynamic performances of SVT [15, 16]. Though they allowed creating a new class of electro-discharged devices (CO_2-lasers and plasmatrons [14]), still there are no theoretical studies of the flow behaviour in SVT based on modern computational fluid dynamics (CFD) simulation technics. The present paper is devoted to the CFD study of strongly swirling gas flows in SVT based on the RSM turbulence model.

[1] Corresponding author: vtvolov@mail.ru

2. Schematics of the problem geometry

Schematics of the problem geometry is shown at figure 1. The basic dimensions of SVT are the following: tube's radius and height are $R_t = 5$ *mm*, $H_t = 11.5$ *mm*, respectively. Central body radius, fillet radius, diffusor radius and diffusor height are $R_{cb} = 0.5$ *mm*, $R_f = 1.5$ *mm*, $R_d = 25$ *mm*, $H_d = 1$ mm, respectively. The width of inlet is $a = 2.1$ *mm* and its height is $b = 1.4$ *mm*.

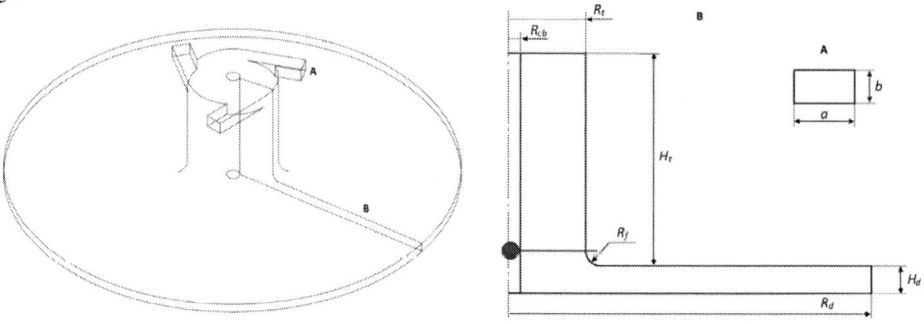

Fig. 1. Schematics of the problem geometry (red dot marks reference frame origin)

3. Mathematical model

The compressible turbulent flows in SVT are governed by mass, momentum and energy conservation equations, together with equation of state. The Reynolds-averaged forms of these equations are:

$$\frac{\partial \rho}{\partial t} + \frac{\partial(\rho u_k)}{\partial x_k} = 0,$$

$$\frac{\partial(\rho u_i)}{\partial t} + \frac{\partial(\rho u_i u_k)}{\partial x_k} = -\frac{\partial p}{\partial x_i} + \frac{\partial}{\partial x_k}\left(\tau_{ki}^{eff}\right),$$

$$\frac{\partial(\rho h_t)}{\partial t} - \frac{\partial p}{\partial t} + \frac{\partial(\rho u_k h_t)}{\partial x_k} = \frac{\partial}{\partial x_k}\left[\lambda\left(\frac{\partial T}{\partial x_k}\right)\right] + \frac{\partial}{\partial x_k}\left(u_i \tau_{ki}^{eff}\right),$$

$$\rho = \frac{Wp}{R_0 T'}$$

where $h_t = h + \frac{1}{2}u_k u_k$ is total enthalpy, $h = C_p T$ is static enthalpy, $\tau_{ij}^{eff} = \tau_{ij} - \rho\overline{u_i' u_j'}$ is the effective stress tensor, $\tau_{ij} = \mu\left(\frac{\partial u_i}{\partial x_j} + \frac{\partial u_j}{\partial x_i}\right) - \frac{2}{3}\frac{\partial u_i}{\partial x_j}\delta_{ij}$ is the viscous stress tensor, W is molar mass, $R_0 = 8.314472\, m^2 \cdot kg/\left(s^2 \cdot K \cdot mol\right)$ is the universal gas constant.

In this study, the flow was considered fully turbulent. For the description of turbulence, Omega Reynolds Stress model was used [16]. It comprises six transport equations for turbulent stresses as well as an additional transport equation for ω, which is similar to the equation of *k*-ω model.

4. Numerical model

For the numerical solution of governing equations, ANSYS CFX commercial CFD software was used. It utilizes a finite volume method. The simulation was carried out with the following boundary conditions: specified values of total (stagnation) pressure and total (stagnation) temperature at inlets, with flow direction at inlets normal to the boundaries; specified values of pressure and temperature at the exit. These values were interpreted as values of static parameters at those parts of boundary where the flow leaves the simulation domain and as values of total parameters at those parts of boundary where flow enters the simulation domain in case of reversed flow. All the walls of SVT were considered adiabatic. The inlet parameters varied: P^* from 0.15 *MPa* to 0.7 *MPa*, T^* from 500*K* to 10,000*K*. The gas was helium, with $W = 4.0$ *g/mol*, $C_p = 5193.0$ *J/(kg·K)*. Transport properties λ and μ were considered as functions of temperature according to [17] for $T^* < 5000K$ and calculated from kinetic theory for $T^* \geq 5000K$.

The calculation was carried out in a pseudo-transient formulation with fixed pseudo-time step equal to 10^{-7} s. Convergence criteria, besides the standard ones based on average normalised residuals, were the global imbalances of mass and energy (the threshold value 0.1%) and constancy of mass flow rate.

The initial computational mesh featured $4.3 \cdot 10^6$ elements ($1.5 \cdot 10^6$ nodes) with the following adaptation, based on pressure gradient, up to $11.3 \cdot 10^6$ elements ($3.0 \cdot 10^6$ nodes).

5. Results and discussion

Comparison of viscous and inviscid flow simulation results revealed that while direct friction and turbulent losses are rather small (from 1 to 10% of power lost due to friction and less than 1% of power lost due to turbulence), viscous effects play a significant role, drastically changing the general flow pattern. Viscous effects are responsible for interaction of "streams" produced by separate inlets. This interaction twists these "streams", turning them into rotating cords (secondary vortices) with high intrinsic swirl (see Figures 2 and 3; results obtained for $T^* = 500K$, $P^* = 0.3MPa$). Such "flow rearrangement" could be considered as a source of energy losses.

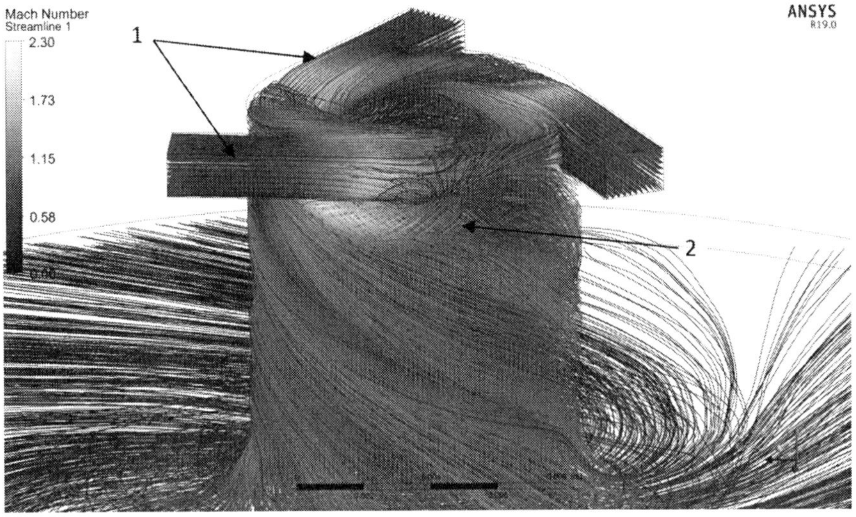

Fig. 2. Streamlines originating from SVT inlets colored by local Mach number: 1 – unswirled initial "streams", 2 – rotating cord (secondary vortex) resulting from their interaction.

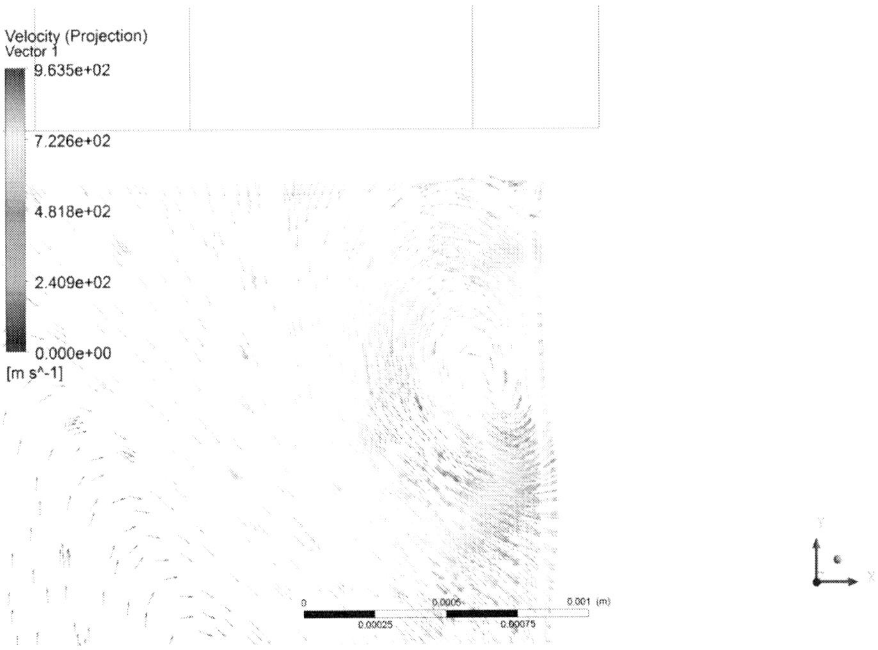

Fig. 3. Cross-section of one of rotating cords formed from inlet "streams" interaction: velocity vectors projected to section plane.

As "streams" become supersonic due to expansion from inlet channels into main chamber, their interaction with each other and with SVT peripheral wall also produces a system of oblique compression shocks. These shocks could be considered as another source of energy losses. Note that viscous effects play a significant role here, too – the pattern of shocks is very different between inviscid and viscous flow (see Figure 4).

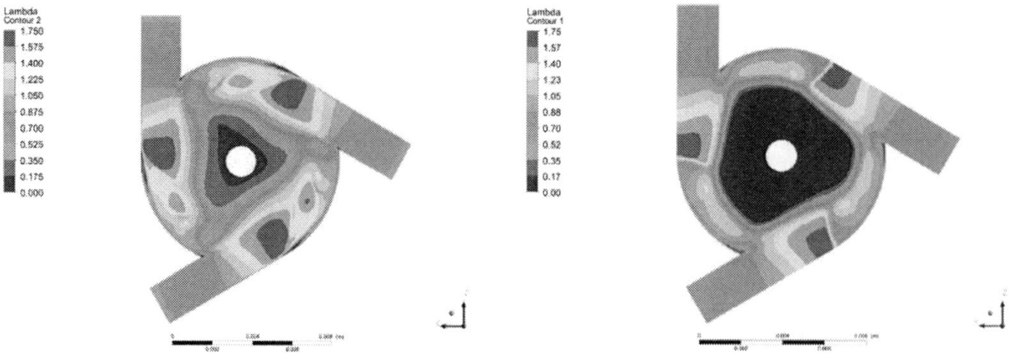

a) Viscous flow b) Inviscid flow

Fig. 4. Velocity coefficient in the inlet plane for viscous and inviscid flow.

Interestingly, it was found that for lower end of the considered temperature range (500 – 10,000K) the main flow could remain supersonic even after passing the shocks (see Figure 5). This was because addition of transonic "relative" velocity (associated with high intrinsic swirl of rotating cords produced by "streams" interaction) to transonic circumferential velocity (associated with general rotation, produced by tangential gas supply) results in supersonic total velocity.

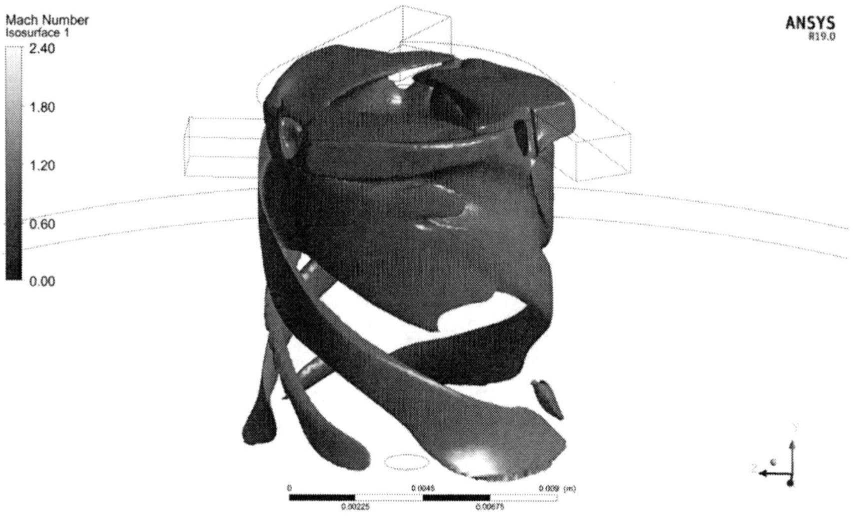

Fig. 5. Isosurface of Mach number equal to 1.

While qualitative analysis of energy losses due to compression shocks presented mere technical difficulties, for qualitative analysis of losses due to "flow rearrangement" and swirl generation it was necessary to estimate the energy associated with the intrinsic swirl of rotating cords. Taking the curl of velocity vector

$\boldsymbol{\omega} = \nabla\times\mathbf{V}$,

and considering it in cylindrical reference frame aligned with SVT main chamber (i.e. axial coordinate z corresponding to Y coordinate of global Cartesian reference frame at Figures 1 and 2) it could be assumed that: its axial component ω_z could be attributed to "general" or "main vortex" rotation, generated by tangential gas supply; its circumferential component ω_θ could be roughly attributed to "secondary" rotation or intrinsic swirl of rotating cords. By analogy with enstrophy, it was proposed to integrate ω_θ^2 by surface at section plane $\theta = const$ to get a crude (by the order of magnitude) estimation of specific energy associated with swirl (note that such a section plane must be truncated in order to omit boundary layer, containing a lot of vorticity). For more precise estimation it was necessary to identify the axis of the rotating cord (to construct the section plane normal to it) together with its bounds and consider the curl of velocity in a local coordinated frame aligned with rotating cord axis.

6. Conclusion

For the first time flow pattern in SVT was studied with state-of-the-art CFD technics. It was found that this pattern is rather complex: it features both subsonic and supersonic regions, systems of oblique shocks, primary rotation (main vortex) and "rotating cords" (secondary vortices) with strong interaction. An attempt was made to estimate the sources of power losses. It was found that only a small amount of losses could be attributed to friction and turbulence (from 2% to 10% due to friction, depending on inlet conditions, with losses due to turbulence making less than 1% for all cases). It was assumed that most of the losses occur due to shocks and secondary swirl generation ("rotating cords" formation).

7. References

[1] Gupta A K, Lilley D G and Syred N 1984 *Swirl Flows* (Turnbridge Wells: Abacus Press) p 475
[2] Novikov N N, Smirnov V A and Mikhailov V V 1988 Vortex burners *Vortex effect and its application in technics* (Kujbyshev: Kuibyshev aviation institute) 92-95
[3] Piralishvili Sh A, Novikov N N and Latyshev A V 1981 Ignition of acetylene in vortex transformer *Vortex effect and its industrial applications* (Kujbyshev: Kuibyshev aviation institute) 132-136
[4] Ranque G J 1933 Experiences sur la détente giratoire avec simultanées d'air chaud et d'un enchappement d'air froid *J. Phys. Radium* **4 (7)** 112–114
[5] Hilsch R 1947 The use of expansion of gases in a centrifugal field as a cooling process *Rev. Sci. Instrum.* **18 (2)** 108–113
[6] Nash J M 1972 The Ranque–Hilsch vortex tube and its application to spacecraft environmental control systems *Dev. Theor. Appl. Mech.* **6** 35–90
[7] Dobratz B M 1964 *Vortex Tubes: a Bibliography* (Livermore: University of California) p 68
[8] Eiamsa-ard S and Promvonge P 2007 Numerical investigation of the thermal separation in a Ranque–Hilsch vortex tube. *Int. J. of Heat and Mass Transf.* **50** 821–832
[9] Farouk T and Farouk B 2007 Large eddy simulations of the flow field and temperature *Int. J. of Heat and Mass Transf.* **50** 4724–35
[10] Gao C M, Bosschaart K J, Zeegers J C H and de Weyl A T A M 2005 Experimental separation on a simple Ranque-Hilsch vortex tube *Cryogenics* **45** 173–187
[11] Merkulov A P 1969 *Vortex Effect and its Applications in Engineering* (Moscow: Mashinostrojenie) p 187
[12] Merkulov A P 1964 Hypothesis of the interaction of vortices *Energetic* **3** 32–37
[13] Merkulov A P and Kolishev N D 1969 Investigation of the vortex tube with diffusor. *Proc. of Kuibyshev aviation institute* **37** 35–51
[14] Volov V T 1983 Method of the vortex diffusor device *Inzhenerno-physicheskiy zhurnal* **1** 35–42.
[15] Volov V T 2006 *Thermodynamic and heat and mass transfer of the strongly twitted gas flows in the energy devices and apparatuses* (Samara: Samara scientific center of RAS) p 316
[16] Wilcox D C 2006 *Turbulence Modeling for CFD* (La Canada: DCW Industries, Inc.) p 522
[17] Green D W and Perry R H 2008 *Perry's Chemical Engineers' Handbook* vol 2 ed Poling B E, Thomson G H et al. (New-York: McGraw-Hill Companies, Inc.) p 517

Mathematical modeling of the impact of forest fires on buildings and structures

Valeriy Perminov*, and *Victoria* Marzaeva

Department of Control and Diagnostics, Tomsk Polytechnic University, 634045 Tomsk, Russia

Abstract. The protection of buildings and structures in a community from destruction by forest fires is a very important concern. This paper addresses the development of a mathematical model for fires in the wildland-urban intermix. The forest fire is a very complicated phenomenon. At present, fire services can forecast the danger rating of, or the specific weather elements relating to, forest fire. There is need to understand and predict forest fire initiation, behavior and impact of fire on the buildings and constructions. This paper's purposes are the improvement of knowledge on the fundamental physical mechanisms that control forest fire behavior. The mathematical modeling of forest fires actions on buildings and structures has been carried out to study the effects of fire intensity and wind speed on possibility of ignition of buildings.

1 Mathematical model

A great deal of work has been done on the theoretical problem of forest fires [1-4]. However, the study of crown fires initiation and spread [9,10] has been limited mainly to cases studied of forest fires propagation without take into account the mutual interaction of crown forest fires with different obstacles (roads, glades and etc.), buildings and constructions. In this paper, the impacts of crown forest fires on building are studied. The dangerous distances between forest and buildings are calculated in cases when the buildings will be ignited under the influence of forest fires. It is assumed that the forest during a forest fire can be modeled as 1) a multi-phase, multistoried, spatially heterogeneous medium; 2) in the fire zone the forest is a porous-dispersed, two-temperature, single-velocity, reactive medium; 3) the forest canopy is supposed to be non – deformed medium (trunks, large branches, small twigs and needles), which affects only the magnitude of the force of resistance in the equation of conservation of momentum in the gas phase, i.e., the medium is assumed to be quasi-solid (almost non-deformable during wind gusts); 4) let there be a so-called "ventilated" forest massif, in which the volume of fractions of condensed forest fuel phases, consisting of dry organic matter, water in liquid state, solid pyrolysis products, and ash, can be neglected compared to the volume fraction of gas phase (components of air and gaseous pyrolysis products); 5) the flow has a developed turbulent nature and molecular transfer is neglected; 6) gaseous phase density doesn't depend on the pressure because of the low velocities of the flow in comparison with the velocity of the

* Corresponding author: perminov@tpu.ru

sound. The modeling approach is based on the use of standard non-stationary three-dimensional conservation. The problem formulated above reduces to the solution of the next system of equations:

$$\frac{\partial \rho}{\partial t}+\frac{\partial}{\partial x_j}(\rho v_j)=Q,\ j=\overline{1,3},\ i=\overline{1,3}; \tag{1}$$

$$\rho\frac{dv_i}{dt}=-\frac{\partial P}{\partial x_i}+\frac{\partial}{\partial x_j}(-\rho\overline{v_i' v_j'})-\rho s c_d v_i|\vec{v}|-\rho g_i-Qv_i; \tag{2}$$

$$\rho c_p\frac{dT}{dt}=\frac{\partial}{\partial x_j}(-\rho c_p\overline{v_j' T'})+q_5R_5-\alpha_v(T-T_s)+k_g(cU_R-4\sigma T^4); \tag{3}$$

$$\rho\frac{dc_\alpha}{dt}=\frac{\partial}{\partial x_j}(-\rho\overline{v_j' c_\alpha'})+R_{5\alpha}-Qc_\alpha,\ \alpha=1,2; \tag{4}$$

$$\frac{\partial}{\partial x_j}\left(\frac{c}{3k}\frac{\partial U_R}{\partial x_j}\right)-kcU_R+4k_S\sigma T_S^4+4k_g\sigma T^4=0, \tag{5}$$
$$k=k_g+k_S;$$

$$\sum_{i=1}^{4}\rho_i c_{pi}\varphi_i\frac{\partial T_S}{\partial t}=q_3R_3-q_2R_2-k_S(cU_R-4\sigma T_S^4)+\alpha_v(T-T_S); \tag{6}$$

$$\rho_1\frac{\partial\varphi_1}{\partial t}=-R_{1s},\rho_2\frac{\partial\varphi_2}{\partial t}=-R_{2s},\rho_3\frac{\partial\varphi_3}{\partial t}=\alpha_C R_{1s}-\frac{M_C}{M_1}R_{3w},\rho_4\frac{\partial\varphi_4}{\partial t}=0; \tag{7}$$

$$\sum_{\alpha=1}^{3}c_\alpha=1,\ P_e=\rho RT\sum_{\alpha=1}^{3}\frac{c_\alpha}{M_\alpha},\ \vec{v}=(v_1,v_2,v_3),\ \vec{g}=(0,0,g).$$

The system of equations (1)–(7) must be solved taking into account the initial and boundary conditions:

$$t=0: v_1=0, v_2=0, v_3=0, T=T_e, c_\alpha=c_{\alpha e}, T_s=T_{se}, \varphi_i=\varphi_{ie}; \tag{8}$$

$$x_1=0: v_1=V, v_2=0, v_3=0, T=T_e, c_\alpha=c_{\alpha e}, -\frac{c}{3k}\frac{\partial U_R}{\partial x_1}+\frac{c}{2}U_R=0; \tag{9}$$

$$x_1=x_{1e}: \frac{\partial v_1}{\partial x_1}=0,\frac{\partial v_2}{\partial x_1}=0,\frac{\partial v_3}{\partial x_1}=0,\frac{\partial T}{\partial x_1}=0,\ \frac{\partial c_\alpha}{\partial x_1}=0,\frac{c}{3k}\frac{\partial U_R}{\partial x_1}+\frac{c}{2}U_R=0; \tag{10}$$

$$x_2=-x_{2e}: \frac{\partial v_1}{\partial x_2}=0,\frac{\partial v_2}{\partial x_2}=0,\frac{\partial v_3}{\partial x_2}=0,\frac{\partial T}{\partial x_2}=0,\ \frac{\partial c_\alpha}{\partial x_2}=0,-\frac{c}{3k}\frac{\partial U_R}{\partial x_2}+\frac{c}{2}U_R=0; \tag{11}$$

$$x_2=x_{2e}: \frac{\partial v_1}{\partial x_2}=0,\frac{\partial v_2}{\partial x_2}=0,\frac{\partial v_3}{\partial x_2}=0,\frac{\partial T}{\partial x_2}=0,\ \frac{\partial c_\alpha}{\partial x_2}=0,\frac{c}{3k}\frac{\partial U_R}{\partial x_2}+\frac{c}{2}U_R=0; \tag{12}$$

$$\begin{gathered}x_3=0: v_1=0, v_2=0,\frac{\partial c_\alpha}{\partial x_3}=0,-\frac{c}{3k}\frac{\partial U_R}{\partial x_3}+\frac{c}{2}U_R=0,\\ \rho v_3=\rho_0\omega_0, T=T_0, |x_1|\le x_0, |x_2|\le x_0,\\ \rho v_3=0, T=T_e, |x_1|>x_0, |x_2|>x_0;\end{gathered} \tag{13}$$

$$x_3=x_{3e}: \frac{\partial v_1}{\partial x_3}=0,\frac{\partial v_2}{\partial x_3}=0,\frac{\partial v_3}{\partial x_3}=0,\frac{\partial T}{\partial x_3}=0,\ \frac{\partial c_\alpha}{\partial x_3}=0,\frac{c}{3k}\frac{\partial U_R}{\partial x_3}+\frac{c}{2}U_R=0. \tag{14}$$

Here and above $\frac{d}{dt}$ is the symbol of the total (substantial) derivative; α_v is the coefficient of phase exchange; ρ - density of gas – dispersed phase, t is time; v_i - the velocity components; T, T_S, - temperatures of gas and solid phases, U_R - density of radiation energy, k - coefficient of radiation attenuation, P - pressure; c_p – constant pressure specific heat of the gas phase, c_{pi}, ρ_i, φ_i – specific heat, density and volume of fraction of condensed phase (1 – dry organic substance, 2 – moisture, 3 – condensed pyrolysis products, 4 – mineral part of forest fuel), R_i – the mass rates of chemical reactions, q_i – thermal effects of chemical reactions; k_g , k_S - radiation absorption coefficients for gas and condensed phases; T_e - the ambient temperature; c_α - mass concentrations of α - component of gas - dispersed medium, index α=1,2,3 where 1 corresponds to the density of oxygen, 2 - to carbon monoxide CO, 3 - to carbon dioxide and inert components of air; R – universal gas constant; M_α , M_C, and M molecular mass of α -components of the gas phase, carbon and air mixture; g is the gravity acceleration; c_d is an empirical coefficient of the resistance of the vegetation, s is the specific surface of the forest fuel in the given forest stratum. To define source terms which characterize inflow (outflow of mass) in a volume unit of the gas-dispersed phase, the formulae were used for the rate of formulation of the gas-dispersed mixture Q, outflow of oxygen R_{51} , changing carbon monoxide R_{52} [3]. It is supposed that the optical properties of a medium are independent of radiation wavelength (the assumption that the medium is "grey"), and the so-called diffusion approximation for radiation flux density were used for a mathematical description of radiation transport during forest fires. To close the system (1)–(7), the components of the tensor of turbulent stresses, and the turbulent heat and mass fluxes are determined using the local-equilibrium model of turbulence [3]. It should be noted that this system of equations describes processes of transfer within the entire region of the forest massif, which includes the space between the underlying surface and the base of the forest canopy, the forest canopy and the space above it, while the appropriate components of the data base are used to calculate the specific properties of the various forest strata and the near-ground layer of atmosphere. This approach substantially simplifies the technology of solving problems of predicting the state of the medium in the fire zone numerically. The thermodynamic, thermophysical and structural characteristics correspond to the forest fuels in the canopy of a different (for example pine [3]) type of forest.

1 Numerical solution and results

The boundary-value problem (1)–(14) is solved numerically. A discrete analog was obtained by means of the control volume method using the SIMPLE algorithm (Patankar [5]). Difference equations that arise in the course of sampling were resolved by the method of SIP [5]. In order to efficiently solve this problem in a reactive flow the method of splitting according to physical processes was used. The accuracy of the program was checked by the method of inserted analytical solutions. The effect of the dimensions of the control volumes on the solution was studied by diminishing them. Fields of temperature, velocity, component mass fractions, and volume fractions of phases were obtained numerically in different instants of time. Fig.1. shows temperature fields at the different instants moments of forest fire spread for a wind speed of 15 m/s (1 – 1.2, 2 - 1.5, 3 – 2., 4 – 3., 5 – 4; $\overline{T}$ =T/T$_e$ T$_e$=300K). The first stage is related to increasing maximum temperature in the place of ignition with the result that a crown fire source appears (Fig.1a). The results of the calculation give an opportunity to consider forest fire spread and distribution of temperature in domain for different instants of time. It is considered the effect of forest fire front on the building which is situated near from the forest on the right side. The results of calculations can be used to evaluate the thermal effects on the building. The temperature fields of crown forest fire at definite moment will be interacted with the obstacle - building (Figure 1 b)-d)) and can be ignited it. During this process, the surface of the wall of the building heats as a result of convection and radiation heat transfer. The wood building will be ignited at definite temperature. It depends on wind velocities, distances from the forest fire to building, the height of building and others parameters.

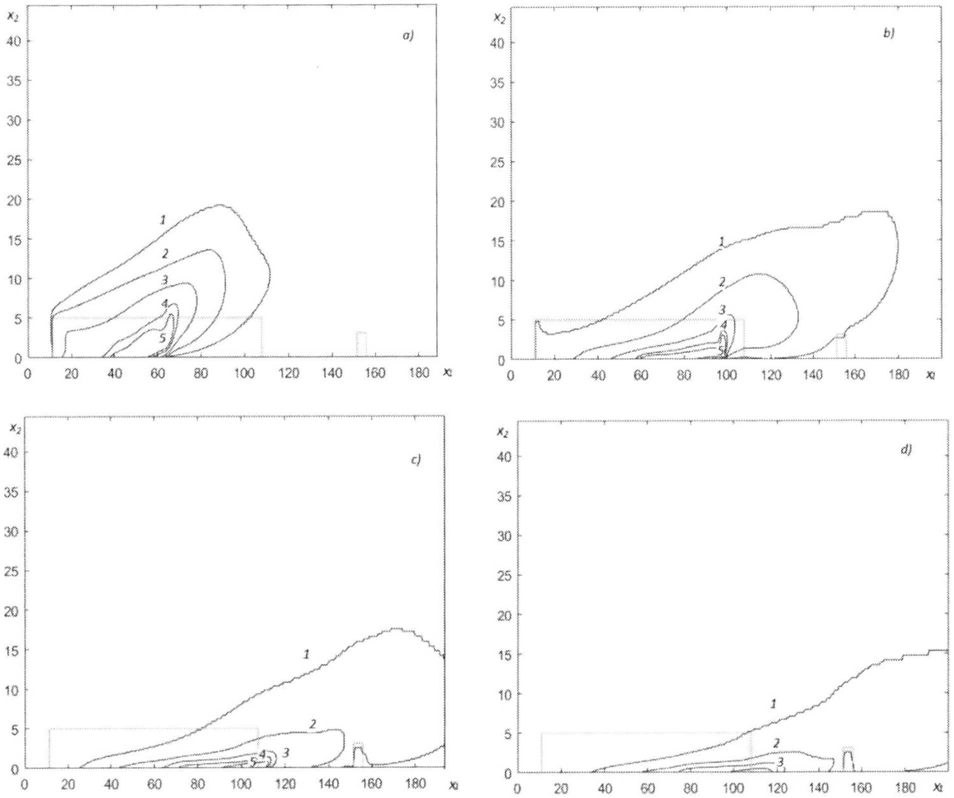

Fig. 1. Gas temperature field at *a) t=15* s, *b) t=20 s, c) t=23 s, d) t=25 s.*

In paper [6] it is showed that the wood will be ignited when its temperature exceeds 300C. As a result of the numerical calculations it was defined maximum safety distances between forest and building when the building would not have been ignited by forest fire (Fig. 2). The wind speed values increase from 6 to 14 m/s. Also, it was studied the influence of the height of building on the value of safety distances. When the height of building changes from 3 to 6 m (line 1 and 2 correspondently), the safety distances *l* also increases (Figure 2) for different values of wind speed.

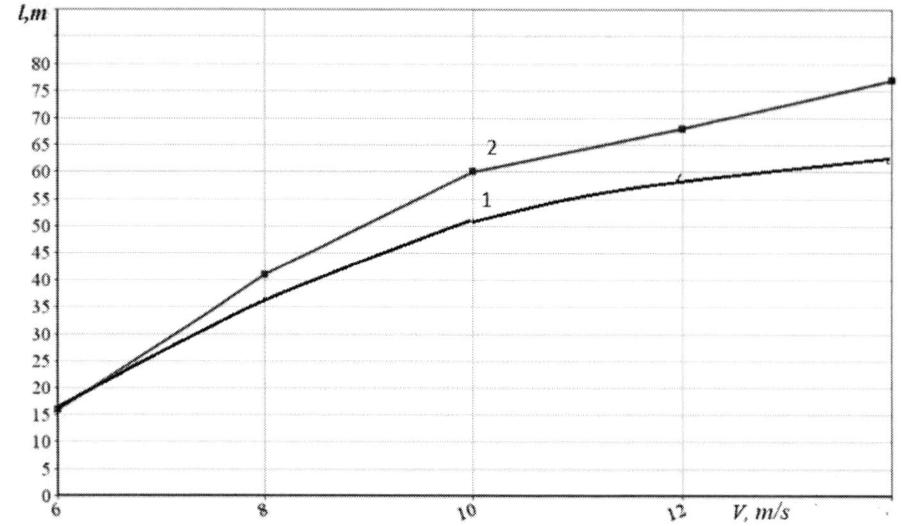

Fig. 2. The dependence of safety distances between forest and building as a function of wind speed values; H=3 m (1) and H=6 m (2).

A multiphase mathematical model of wind-aided crown forest fires propagating through heterogeneous fuel beds has been performed. It takes into account the hydrodynamic aspects of the flow and uses Arrhenius kinetics to describe the basic physics and chemical processes of thermal decomposition heating, drying, pyrolysis, and combustion. Turbulence and radiation are considered in order to improve the physical insight. It allows to investigate the dynamics of the impact of forest fires on buildings under the influence of various external conditions: a) meteorology conditions (air temperature, wind velocity etc.), b) type (various kinds of forest combustible materials) and their state (load, moisture etc.). The calculations let to get the maximum distance from the fire to the building in which the object possible ignition. It has been found that the effect of increasing the wind speed is to increase the safety distances between forest and building. The increasing of building height is observed also led to increase the safety distances between forest and building. Specific experiments are also needed to obtain more reliable information on validation of further solution of this problem.

The paper was supported from RFBR (project code: № 16-41-700022 p_a) and within the framework of Tomsk Polytechnic University Competitiveness Enhancement Program grant.

References

[1] R.C. Rothermal, Predicting behavior and size of crown fires in the Northern Rocky Mountains, Res.Pap. INT-438. Ogden, UT: US Department of Agriculture, Forest Service, Intermountain Forest and Range Experiment Station (1972)
[2] F.A. Albini, F.A. et al., Int. J. of Wildland Fire, **5** (1995)
[3] A.M. Grishin, Mathematical Modeling Forest Fire and New Methods Fighting Them. Publishing House of Tomsk University, Tomsk (1997)
[4] D. Morvan, J.L. Dupuy, Combustion and Flame **138** (2004)
[5] S.V. Patankar, Numerical Heat Transfer and Fluid Flow, Hemisphere Publishing Corporation, New York (1981)
[6] E.N. Valendik, P.M. Mathveev, M.A. Safromov, Large Forest Fires and Fighting with Them. Science, Moscow (in Russian) (1979)

Refining the workflow of combustion chamber with toroidal recirculation mixing zone

Mikhail Orlov[1], *Vladislav* Anisimov[1]*, *Oleg* Kolomzarov[1] and *Mikhail* Anisimov[1]

Samara National Research University, 443086, Moskovskoye Shosse 34, Samara, Russia

Abstract. The object of the study is a straight-flow annular combustion chamber (CC) with toroidal recirculation-mixing zone which has significant reserves of minimization in size. The workflow of the CC has a number of problems, which arise due to the lack of experimental and calculated data. The aim of the study is to refine the design of the CC with toroidal recirculation-mixing to obtain the necessary workflow parameters. The first step of maturation involved the organization of swirl structure at primary zone. On the second step there were conducted the refining of the flame shape in primary combustion zone. The third step was devoted to the formation of the necessary temperature field at the outlet of the CC. As a result, an acceptable design of the combustion chamber was obtained.

1 Introduction

The design of the serial power plant was used as a prototype to create a new engine. In a new project the scheme solution of inlet duct and centrifugal compressor was not changed, but radial turbine became axial. This entailed a change in the combustion chamber. After working out a number of possible schemes, it was decided to choose a straight-flow combustion chamber (CC) of annular type with a toroidal recirculation-mixing zone, which has substantial reserves of minimization in size. A feature of this combustion chamber is its diagonal arrangement relative to the axis of the engine. There were several problems with the CC workflow, which are related to the lack of experimental and calculated data [1-6].

The aim of the paper is the maturation of the design of the combustion chamber with toroidal recirculation mixing zone for optimization its workflow.

To achieve this goal, it is necessary to solve the following tasks:

1) refine the aerodynamic flow structure to form a toroidal vortex in the primary combustion zone;

2) to form the combustion process in such a way that there is no local overheating of the flame tube wall;

3) provide the required temperature field at the outlet from the combustion chamber.

[1] Corresponding author: vradik@mail.ru

2 Methods

The maturation of CC was carried out by simulation method in software package ANSYS Fluent. The scheme of the initial design of the combustion chamber is shown in Figure 1. Data of the mode parameters which are necessary to perform the simulation were taken from the results of the thermogasodynamic calculation.

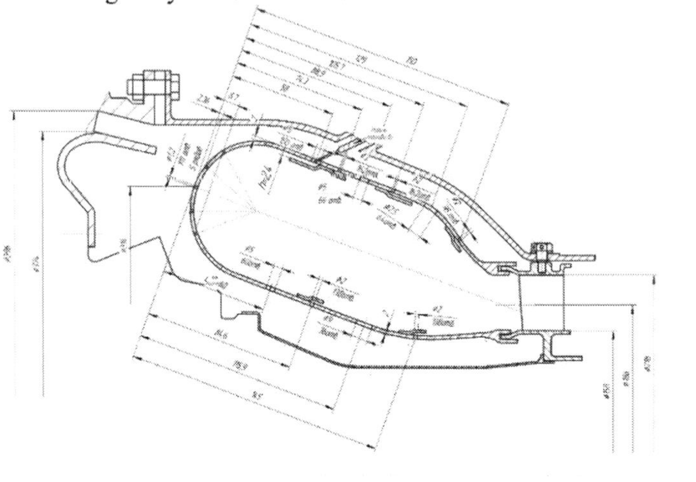

Fig. 1. The scheme of basic version of CC.

In the beginning, the sector of volume geometrical model of the basic version of CC was created (fig. 2, a). This sector corresponds to 120° and includes four injectors.

On the basis of the geometric model, a grid of finite elements was generated (fig. 2, b). The number of elements of the grid model is 12.5 million. The skewness is 0.87, which indicates the acceptable quality of the grid. A periodic condition was set on the side faces of the sector.

The choice of simulation models was determined by the accuracy of modeling the workflow in the combustion chamber [7, 8]. The following calculation models were chosen: turbulent model – Reynolds Stress, radiation model – Discrete Ordinates, combustion model – Non-premixed, kinetic mechanism – Jeta 39. Kerosene was chosen as a fuel. The calculation was carried out in a steady state for the nominal operating mode of the engine. The temperature and total pressure of the air flow at the inlet were set as the boundary conditions, as well as the total pressure and gas flow rate at the outlet.

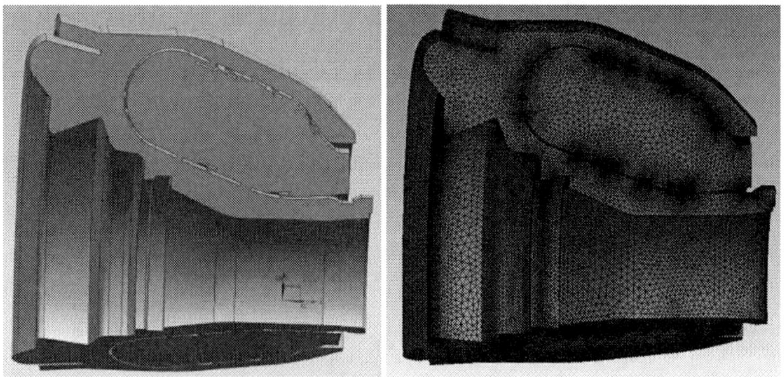

a b

Fig. 2. The volume geometrical (a) and mesh (b) models of CC.

3 Refining

3.1 The first stage

At the first stage, the refining of the flame tube design was carried out to organize the vortex structure in the primary zone.

On the vector field (fig. 3, a) obtained from simulation in the cross section of the nozzle, it can be seen that the design of the initial version does not allow us to give the required swirled flow. This results in insufficient mixing of the flow and afterburning of the mixture in the dilution and mixing zones. To obtain a stable and developed recirculation zone, it was suggested to use a protuberance on the inner part of the flame tube that performs the function of a "springboard". Due to the insufficient swirling of the flow in the primary zone, additional changes were made:

- the air supply holes were offset by 10 mm downstream, to increase the mixing path and evaporation of the fuel-air mixture.
- the angle of the holes on the front wall was changed. It was changed from the normal direction of the wall to 45 degrees to the normal, to create an air curtain and reduce the resistance to the flow that runs along the wall.
- the nozzle was moved to the air intake hole of the combustion zone into the gap between the flame tube and the deflector to increase the time for mixing the fuel and air and its evaporation.
- a protrusion was made on the inner part of the flame tube to direct the flow and its subsequent swirl.
- two rows of cooling holes were retained on the flame tube wall near the primary zone, but their diameters were reduced and the number was increased to preserve the total area of the holes on the wall of the toroidal part.

The result of the simulation obtained for the design after the introduction of the measures described above is shown in Figure 3. It clearly shows the formed structure of the stream recirculation in the primary combustion zone.

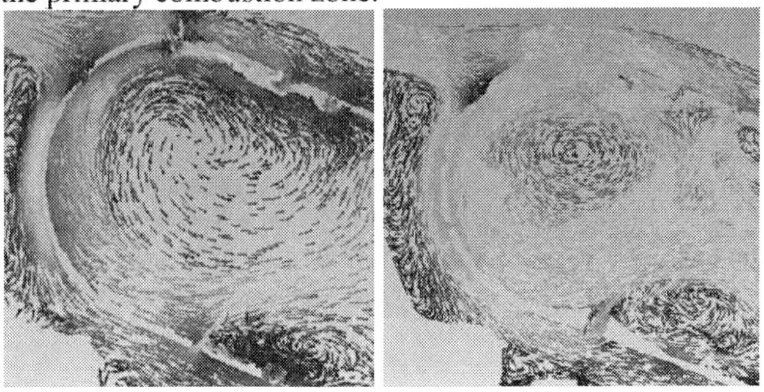

a b

Fig. 3. Velocity vector field of the basic version (a) and after the first step (b).

3.2 The second stage

On the second step, the refining of the flame form in the primary zone was carried out. This zone should have sufficient swirled flow and should stabilize the flame within the recirculation-mixing zone. It will allow the fuel to burn almost completely in the primary zone and only a small amount of it will burn downstream. The original injectors were replaced with the jet nozzles with a spray angle of 12..15° to form a stable and effective combustion zone. Blade vortexers were added in the gap between the deflector of the air intake holes and the flame tube in order to improve the mixing of fuel and air. Figure 4 shows the comparison of the temperature fields in the primary zone of basic version (fig. 4, a), after the first step of maturation (fig. 4, b) and after the second step (fig. 4, c).

When comparing the first two fields, it is seen that the insufficient formation of a swirl in the primary zone leads to the burning of the fuel in the downstream zones and, as a consequence, the combustible mixture is brought to the outer wall of the flame tube. This can cause its overheating and burnout.

After the second step of maturation the mixing of the mixture was improved and the flame was concentrate in the recirculation-mixing zone. The small flame areas appeared outside the primary zone due to stabilization behind rows of large holes, but the slope of the cooling holes allowed reducing the thermal load on the wall without increasing the resistance to the main flow (fig. 4, c).

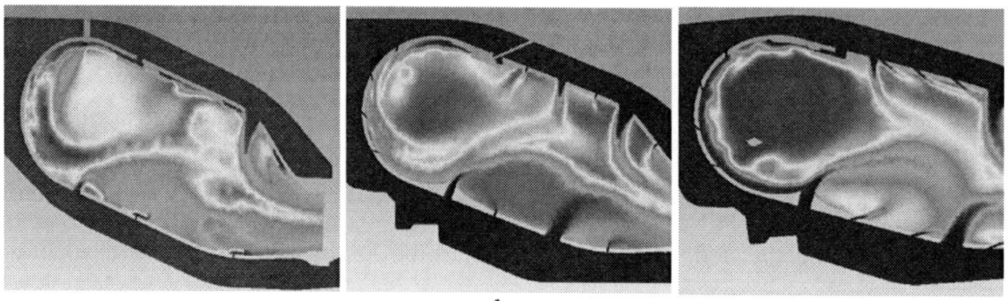

a b c

Fig. 4. The temperature fields at the cross section of injector of basic version (a), after the first step (b) and after the second step (c) of maturation.

3.3 The third stage

The third stage includes the formation of necessary temperature field at the outlet of CC. To reach the goal, the following events were carried out: make the necessary depth of penetration of air jets, change the number and location of the mixing holes.

As a result of these design changes, it was possible to obtain an acceptable temperature field distribution at the outlet of the CC (version 3.4, fig. 5, a). Since the change in the parameters forming the mixing zone is inherently associated with the air distribution in other areas of the CC, when the area of the holes in the mixing zone was changed, the areas of the holes in the other zones were corrected. Figure 5 (b) shows the temperature field at the outlet of CC. As can be seen, the final version 3.4 does not have overheating zones from flame tongues at the outlet, which partially exist at other variants.

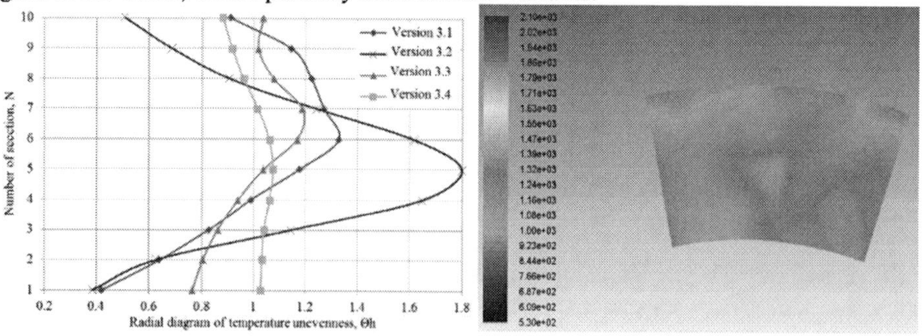

a b

Fig. 5. Radial diagram of temperature unevenness (a) and the temperature field (b) at the outlet of CC.

4 Discussions

Thus, it can be stated that as a result of the refining, an acceptable design of the combustion chamber was obtained.

The simulation methods allowed us to make the refining of CC design, organize the vortex structure and necessary flame form in the primary zone and temperature field at the outlet. Also, the use of computer simulation for the design refining has accelerated the process of getting a new CC design by about five times in comparison with the existing experience of traditional methods.

Based on the results of the work, the pilot sample of the CC was created by the methods of prototyping (laser sintering). The production cycle is reduced by at least 10 times compared with the use of traditional technologies.

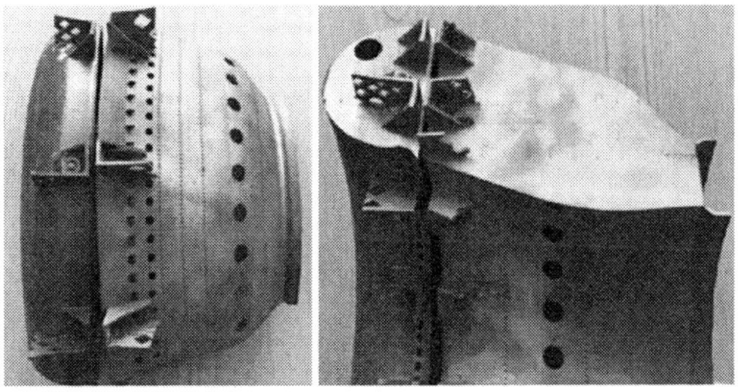

Fig. 6. The flame tube of developed combustion chamber.

Continuation of work occurs in the fire tests of the developed combustion chamber.

This work was supported by the Ministry of education and science of the Russian Federation in the framework of the implementation of the Program "Research and development on priority directions of scientific-technological complex of Russia for 2014-2020" (RFMEFI58716X0033).

References

1. A.M. Lansky, S.V. Lukachev, O.V. Kolomzarov. Vestnik Moskovskogo Aviatsionnogo Instituta, **23**, 3, 47-57 (2016)
2. V.Y. Abrashkin Izvestiya Samarskogo nauchnogo tsentra Rossiyskoy akademii nauk, **8**, 4, 1136-1141 (2006)
3. V.P. Lukachev, A.M. Lansky, V.Y. Abrashkin, A.A. Didenko, P.G. Zubkov, Y.L. Kovilov, S.G. Matveev, A.M. Tsiganov, M.A. Shamban, V.A. Yakovlev, Vestnik SGAU, 1, 11-39 (1999)
4. E. Carr, Futher applications of the lucas fan spray fuel injection combustion system // ASME, 1985. – P. 1-8.
5. E. Carr, H. Todd, The design and performance of a reverse flow combustion system for the TP 500 gas turbine engine, ASME, 1-6 (1989).
6. Carr E. The combustion of a range of distillate fuels in small gas turbine engines, ASME, 1-9 (1979).
7. M.Y. Orlov, V.M. Anisimov. Vestnik Moskovskogo Aviatsionnogo Instituta, **24**, 2, 50-56 (2017)
8. V.M. Anisimov, M.Y. Orlov, I.A. Zubrilin. Vestnik Moskovskogo Aviatsionnogo Instituta, **23**, 3, 29-39 (2016)

Results of Numerical Modeling of Combustion Processes in a Vortex Chamber

Artem V. Badernikov[1,*], *Shota* A. Piralishvily[2], and *Alexander* I. Guryanov[2]

[1]UEC-Saturn, Rybinsk
[2]Rybinsk State Aviation Technical University

Abstract. The results of numerical modelling of combustion in a vortex chamber are presented. The calculations are performed using k-ε and k-ε with curvature correction of streamlines turbulence models. Upon combustion calculation of the well mixed mixture the Burning Velocity Model (BVM) was applied. The usage of the streamlines curvature correction allows to increase the accuracy of the prediction of the circumferential velocity component and the temperature field in the vortex chamber.

1 Introduction

Investigations of the oppositely directed swirling jets interaction confirm the presence of shear mixing layers that generate high-intensity turbulence, large-scale eddies and three-dimensional vortex structures. The observed effects are used to organize intensive mixing of fuel and air components, followed by firing augmentation of the prepared fuel-air mixture with high intensity in a limited volume. The mentioned concept has developed in the designs of various reverse-flow burner units, the basis of which is the modified design of the vortex tube [1].

The purpose of this paper is to investigate the usage possibility of the correction model for the streamlines curvature in calculation of highly swirling counter-directed flows, and also to determinate the main modelling factors which influence the solution.

2 Experimental part

The investigation of spatial distribution of mean tangential velocity, mean temperature and mean axial velocity profiles inside and at the outlet of the vortex chamber, was made in the paper [2]. The vortex chamber with the diameter D_{ch} = 152.5 mm and relative dimensions ($\bar{F}_{inlet}$= 0.12, $\bar{R}_{outlet}$= 0.511, $\bar{L}_{ch}$= 2) is equipped with a quartz glass window for laser-optical measurements of the flow velocity and holes for platinum-rhodium thermocouples. The premixed mixture of air and natural gas was used as a fuel. Simplicity of configuration and a sufficiently large amount of experimental data in the paper [2] make it possible to use it as a model task for adjusting the physical and mathematical model that correctly reflects

* Corresponding author: badartem@mail.ru

the characteristic features inherent in the swirling flow in a vortex reverse-flow burner module, taking into account the combustion reactions.

3 Calculation model

Geometry for the 3D CFD simulation and calculation model of the vortex chamber is shown in Fig 1 and Fig. 2 respectively. For calculation purposes, the initial geometric model of the vortex chamber was simplified. Also to save the computer resources, the full-scale model was replaced by sector 1/8, on the lateral boundaries of which the boundary conditions of periodicity were set. To increase the stability of the calculation, a volume simulating the atmosphere condition was added to the exit from the vortex chamber. The rest of the geometry was modelled without any simplification.

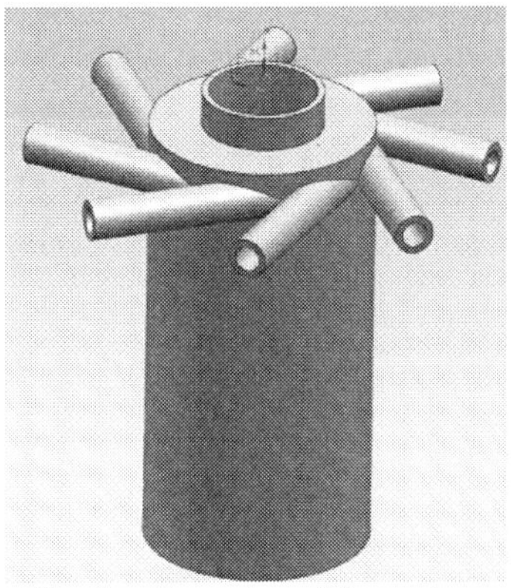

Fig. 1 – Geometry of the vortex chamber

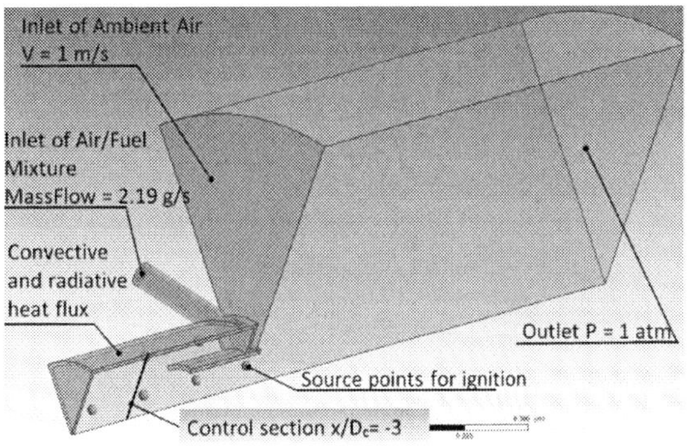

Fig. 2 - Vortex chamber calculation model

The boundary conditions simulated the vortex chamber testing mode: Premixed mixture of fuel and air with an excess air factor $\alpha = 1.1$ and flow rate $G = 2.19$ g/s (per sector 1/8) was applied to the tangentially located ports. The Reynolds number $Re = V_{in}*R_{ch}/\nu$ was $Re = 34400$; output pressure was $P = 101325$ Pa. To ensure the stability of the calculation, the boundary condition of the inlet with airflow velocity of 1 m/s was established from the front side of the free volume. On the external surface of the vortex chamber, the boundary conditions of the heat flux were set, which model the heat exchange with the ambient environment: the convective part of the heat flux was calculated according to the formula $q_{conv} = \alpha\ (T_{wall}-T_{air})$, where $\alpha = 5$ W/m^{2}K; the heat flux from the radiation was calculated using the formula $q_{rad} = \sigma{\cdot}\varepsilon{\cdot}[(T_{wall})^4-(T_{air})^4]$, where σ=5,67E-8 – the Stefan-Boltzmann constant, ε=0,8 – emissivity of the wall.

The calculations of the reacting flow were made using both the original turbulence k-ε model, and k-ε with streamlines curvature correction (k-ε CC) [3]. For the calculations, the Burning Velocity Model (BVM) was used [4], the chemical reaction kinetics was modelled by detailed kinetic mechanism containing 28 substances and 100 reactions.

The calculations were made by ANSYS CFX. For the calculation purposes, a block-structured hexahedral mesh with 300412 elements was built. The quantity of elements in the axial, radial and circumferential directions was chosen according to the basis of the previous study of the mesh convergence for the problem of air flow in a cyclone chamber, and was 150x45x30, respectively.

3.1 Model of streamlines curvature correction for two-parameter turbulence models

The two-parameters turbulence k-ε model is used for modelling a wide range of tasks, it's strengths are simplicity, good convergence of calculation, universality, while maintaining good accuracy. The known shortcomings of this model are manifested in not quite accurate modelling: Points of flow separation from smooth surfaces, flows in the boundary layer, strongly swirling flows, and very curved streamlines.

Some of the drawbacks of vortex viscosity models is that they are insensitive to curvature of flow lines and the rotation of the system. To take these effects into account, in the paper published by Spalart and Schur [3], the modification of the turbulent pulsations generation term is suggested, which increases the two-parameter models sensitivity to these effects.

The effect of the streamlines curvature correction leads to vortex viscosity decrease, which, in case of highly curved flows, makes it possible to realize the flow motion according to the free vortex law and to obtain the results which are adequate to the experiment. It should be noted that in case of the hydrocyclone calculation, the SST-CC model with the streamlines curvature correction shows the results close to the RSM Reynolds stress model, taking into account turbulence anisotropy, while it is much more saving in terms of the computing resources [5].

4 Results

The position of the flame front recorded experimentally is shown in Fig. 3, and the isolines of the temperature field obtained from the experiment is shown in Fig. 4.

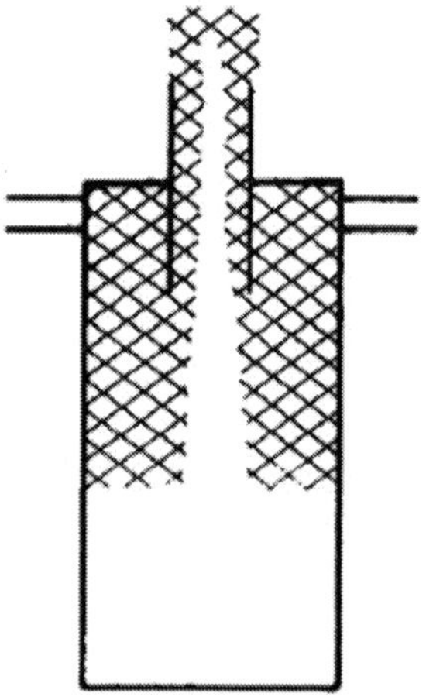

Fig. 3. – Schematic picture of the flame front position (experimental data [2])

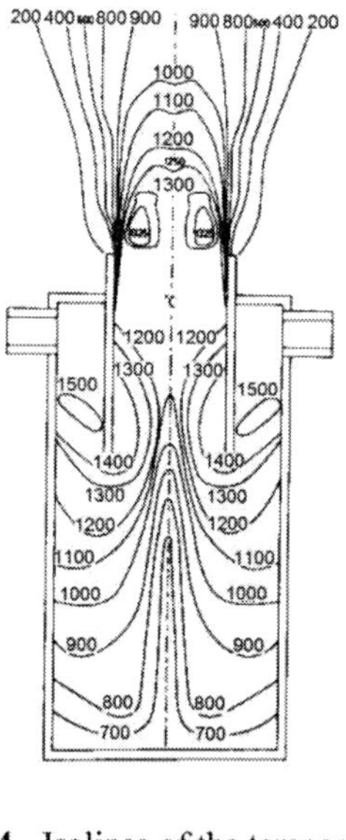

Fig. 4.- Isolines of the temperature field (experimental data [2])

4.1 Influence of a nonadiabatic wall

To determinate the extent of how much the heat transfer of combustion products influence the walls of the vortex chamber, the calculations were made by using two scenarios:

Assuming that the walls are adiabatic (no heat exchange with environment) in one case, and non-adiabatic walls of the vortex chamber (convective and radiation heat flux was taken into account) in the other. The calculation results are shown in Fig. 5 and 6.

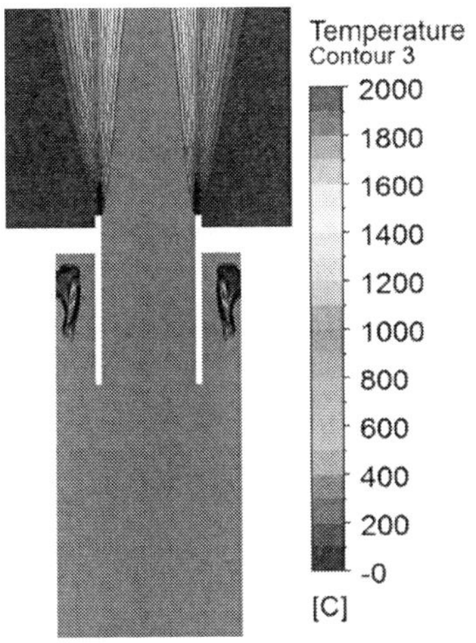

Fig. 5. - Temperature field. Calculation with adiabatic walls

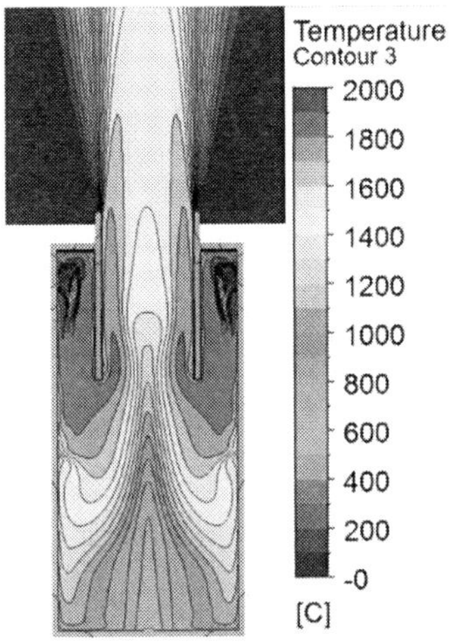

Fig. 6. - Temperature field. Calculation with non-adiabatic walls and k-ε CC

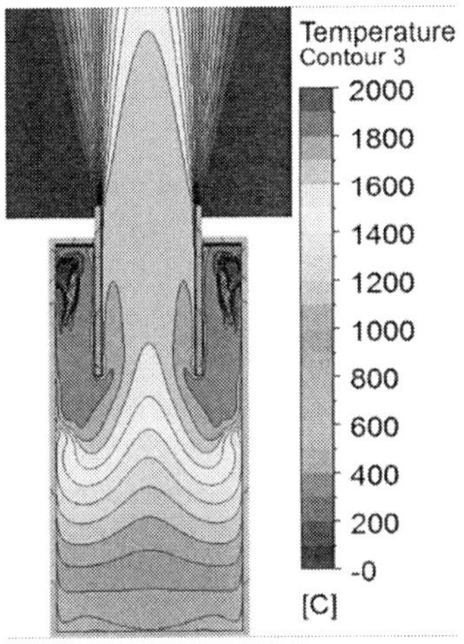

Fig. 7. - Temperature field. Results of calculations using the model k-ε without modifications.

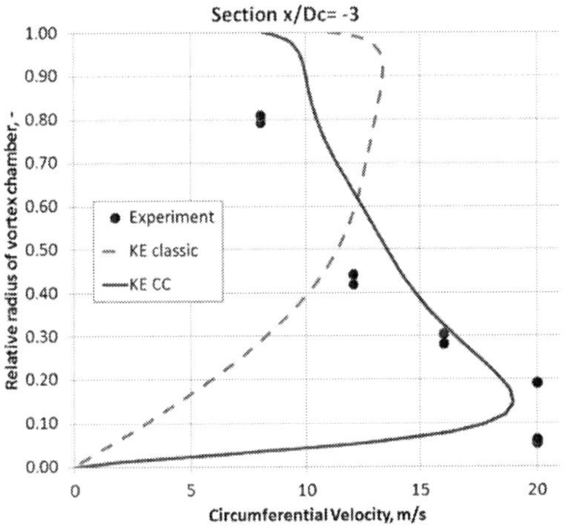

Fig. 8. - Circumferential velocity component in the cross section x/Dc = -3.

As you can see in the figures, taking into account the non-adiabaticity of the vortex chamber walls plays the significant role in the formation of the temperature field in the rear part of the vortex chamber and the near-axis zone. Due to the vortex flow, the combustion products stay in the vortex chamber for a long time, and give off the heat to the walls, while the flow is directed from the periphery to the axis. This forms the region of the cooled combustion products in the near-axis zone, which was noted in the experiment.

Thus, for qualitative modelling of the temperature field in the vortex chamber, it is necessary to take into account the heat exchange with the walls, so further calculations are made with non-adiabatic walls.

4.2 The effect of the curvature correction of the streamlines

It is interesting to compare the results obtained using the classical turbulence k-ε model without modifications and with modification which takes into account the curvature of the flow lines. The calculation results which were made using the k-ε model without modifications are shown in Fig. 7; with streamlines curvature correction - in Fig. 6. Fig. 8 shows the graphs of the circumferential velocity component.

According to the Fig. 8 and 6 the usage of the streamlines curvature correction allows to increase the accuracy of the prediction of the circumferential velocity component and the temperature field in the vortex chamber.

The analysis of the figures 7 and 6 shows that the variant of the calculation in which the turbulence k-ε model is used without modifications, the combustion products are more intensively mixed in the vortex chamber area vs. the variant with the curvature correction. Temperature in the near-axial region of the vortex chamber is thus increased vs. the experiment.

5 Conclusions

1) For qualitative modelling of the temperature field in the vortex chamber, it is necessary to take into account the heat exchange with the walls.

2) The usage of the streamlines curvature correction allows to increase the accuracy of the prediction of the circumferential velocity component and the temperature field in the vortex chamber.

For further calculations of the vortex counter-flow burner modules, it is expedient to use the turbulence k-ε model in conjunction with the streamlines curvature correction model and the BVM combustion model. For qualitative modelling of the temperature field in the vortex chamber, it is necessary to take into account the heat exchange with the walls.

6 Acknowledgments

This work was financially supported by RFBR (Russian Foundation for Basic Research) grant No. 16- 08-00974 "Development of a model description of heat and mass exchange processes within swirling limited flows and its application in the development of advanced technology devices."

7 References

1. V.V. Biryuk, *Vortex effect (Technical applications). Volume 2 (Part 1)* / V.V. Biryuk, S.V. Veretennikov, A.I. Guryanov, Sh.A. Piralishvily, LLC "Nauchtekhlitizdat" pp. 177-272 (2014)
2. S.E. Najim, A.C. Styles, N. Syred, Flame movement mechanisms and characteristics of gas fired cyclone combustors/ *Eighteen Symposium (International) on Combustion. The Combustion Institute* (1981)
3. P.R. Spalart, M. Shur, On the sensitization of turbulence models to rotation and curvature, Aerospace Sci. Tech., **1(5)**, pp. 297-302, (1997).
4. ANSYS CFX-Solver Modelling Guide. / ANSYS Inc, (2016).
5. P.E. Smirnov, F.R. Menter, "Sensitization of the SST turbulence model to rotation and curvature by applying the Spalart-Shur correction term", *ASME Paper GT 2008-50480*, (2008).

Percolation model of combustion

D. Rybakov[1,*], *Kh. Lamazhapov*[1]
[1]Samara state transport university, Russian Federation

Abstract. Strong dependencies in equations cause dramatic boost of reaction rates in some rare but meaningful areas of a combustion gap. That leads to intermittent behavior of some values and for these reasons averaging approaches in calculations are not relevant. We use percolation theory instead in order to explain ignition delay and completeness of combustion.

In ideal conditions reaction rate depends on molecular concentration and some factor k as

$$-da_1/dt = -da_2/dt = k\,a_1a_2 \tag{1}$$

Here the factor k depends on temperature as

$$k=k_0e^{-ER^{-1}T^{-1}} \tag{2}$$

and apparently this coefficient has a remarkable variation range. Assume typical activation energy E equals 40 $kcal\cdot mol^{-1}$, the universal gas constant R equals 1.99 $kcal\cdot mol^{-1}\cdot K^{-1}$ and temperature T ranges over 500 to 1000 Kelvin. Then here we observe $k(500)=10^{-17.4}$ and $k(1000)=10^{-8.7}$ so

$$k(1000)/k(500)\approx 10^{-8.7} \tag{3}$$

So if temperature T increases 2 times the coefficient k(T) will increase billion times! [1]

Molecular concentrations a_1,a_2, temperature T and other magnitudes are statistically distributed in real life. Also, temperature has a feedback coupling with local energy output. Therefore, values form random fields in a combustion gap. Stochastic behavior leads to dramatic intensification of combustion processes in some rare but meaningful areas. Moreover, one can investigate intermittent random fields using statistical moments applying the following rule: if M_i increases as i increases this will point out irrelevancy of averaging approaches [2].

$$M_i=\sqrt[i]{\int_0^{\infty}(k(x))^i\cdot\frac{e^{-(\mu-x)^2/2\sigma^2}}{\sigma\sqrt{2\pi}}dx} \tag{4}$$

The root sign is used in order to keep physical unit consistency. Numerical calculation for T~N(μ=100, σ=100) yields the following moments: $M_1=8\cdot10^{-9}$, $M_6=8\cdot10^{-7}$, $M_{12}=8\cdot10^{-6}$.

For instance, a fuel spray notably clear demonstrates intermittent properties because it definitely encapsulates pure fuel droplets and areas of irregularly saturated vapor. There exists a set of small areas giving a boost to local burning process. These areas configure clusters with very irregular structure and could be described by fractal dimension D. So in

* Corresponding author: dim2r@yandex.ru

some cases inflaming front should pass a very long path in between different areas with very different properties. So some physical properties directly depend on irregular geometry of different areas. And geometry depends on active element volume fraction ρ. In our article, value ρ takes into account element intersections and represents final volume fraction. Critical value could be obtained analytically as well as numerically ρ_c=0.29[4]. We performed 3D computer simulation over percolation system with soft-core spherical elements in order to find out some properties and features of clusters and surfaces important for combustion theory (figure 1). Model includes a set of randomly located spheres such that a single sphere represents a fuel droplet surrounded by fuel vapor. We assume that ignition and flame propagation is possible only within spheres and clusters.

Fig. 1. Different representations of the same scene including a droplet cluster. The cluster is red. The path inside the cluster is green. Orphan droplets are black.

Computer simulation shows that flame propagation time $t_{prop}(\rho)$ from one point to another depends on the shortest path length between two points $L_{path}(\rho)$. The length significantly increases at the percolation threshold and exceeds size of a gap L. So lengths of paths explain an ignition timeout and its statistical properties (figure 2).

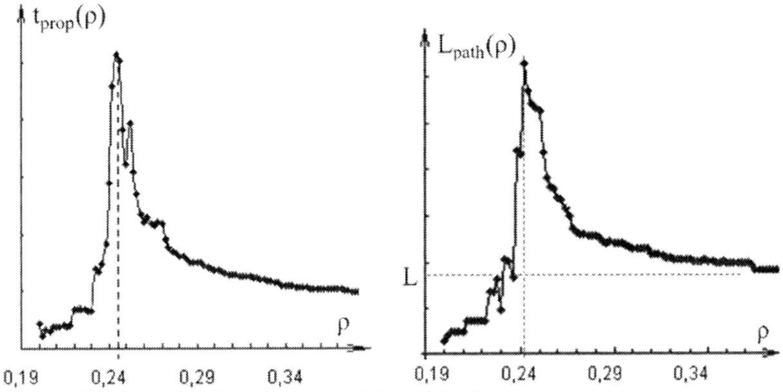

Fig. 2. Flame propagation time vs path length inside a cluster.

At the beginning stage a flame propagates beneath a cluster surface. The closer elements located the smaller total surface is available from outside a cluster. We calculated available surface fraction as

$$F(\rho)=\sum s_i/\sum S_i \tag{5}$$

Here $S_i=4\pi r^2$ is total element surface and s_i is external fraction of element surface. Ball size variation within an ensemble has moderate influence on external surface fraction (figure 2).

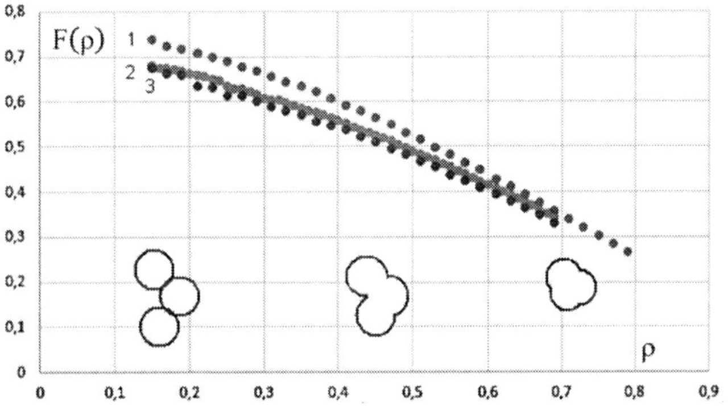

Fig. 3. Cluster surface relative to total surface of elements. 1) Balls have equal size.
2) Ball radiuses are statistically distributed $R_{max}/R_{min}=5$
3) Ball radiuses are statistically distributed $R_{max}/R_{min}=10$

Also in some cases path does not exist between two points. That means that some areas are inaccessible by the flame. So completeness of combustion C is attributable to a balance of accessible and inaccessible areas in a percolation system.

$$C(\rho)=N_{cluster}/N_{total} \tag{6}$$

At the best case ignition starts from an element belonging to a biggest cluster. Thus completeness of combustion does not exceed a fraction of the biggest cluster in the whole system (figure 4).

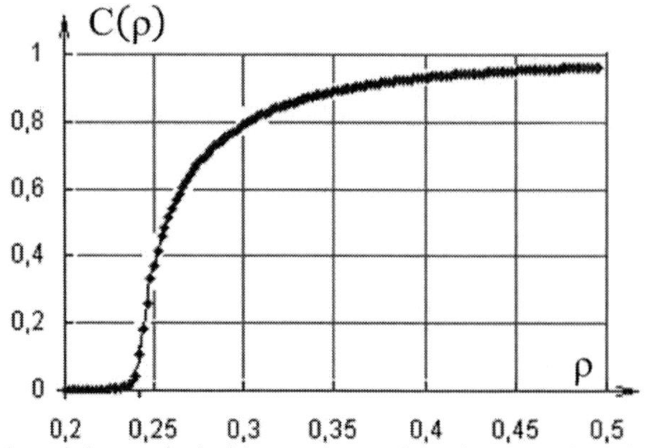

Fig. 4. Relative number of element in the biggest cluster. Completeness of combustion at the initial stage does not exceed that fraction.

References

1. Ya. B. Zel'dovich, G.I. Barenblatt, V.B. Librovich, G.M.Mahviladze, *Mathematical theory of combustion and explosion*. Science, (1980)
2. Ya. B. Zel'dovich, S. A. Molchanov, A. A. Ruzmaikin, and D. D. Sokolov, *Intermittency in random media*. Sov. Phys. Usp., (1987)
3. M. Sahimi, *Applications of percolation theory*. London : Taylor & Francis, (1992)
4. I. Balberg, C. H. Anderson, S. Alexander and N. Wagner, *Excluded volume and its relation to the onset of percolation*. Phys. Rev. B, **30,** 7, (1984)

MATEC Web of Conferences **209**, 00025 (2018) https://doi.org/10.1051/matecconf/201820900025
ComPhysChem'18

Experimental investigation of a gel fuel combustion process initiated by a hot particle

Aleksandr Nigay*, *Dmitriy* Glushkov

Heat and Mass Transfer Simulation Laboratory, National Research Tomsk Polytechnic University, 30 Lenin Avenue, Tomsk, 634050, Russia

Abstract. Experimental studies were performed for crude oil-based fuel samples. The initial temperature of the samples varied down to 120 K. Ignition was performed by single metal particles of various shapes and temperatures, which reached 1350 K. A specially developed experimental setup allowed recording of the proceeding processes at a high speed. As a result, the characteristics of physical and chemical processes were analysed. Conditions necessary for stable ignition and ignition delay times were determined depending on various conditions.

1 Introduction

In recent years, the urgent task is to develop new fuels and ways to initiate their combustion [1]. It is difficult to perform measurements of ignition and combustion processes because of their high intensity. Therefore, theoretical study in the near future can be the basis for further R&D (research and development). To conduct such studies we need adequate mathematical models of physical and chemical processes. They should be based on the initial data in the form of thermophysical characteristics and kinetic parameters of the processes. Information about this problem was analysed in international literature [2-4], for example. Studies of the heat and mass transfer processes and chemical reactions during the ignition of gel fuels have not been carried out. Thus, we will study these processes and reactions both experimentally and theoretically.

Gel fuels have higher energy characteristics in comparison with composite solid propellants, in particular the specific impulse of 300–350 seconds at a force of 1 kg per 1 kg of fuel [5]. Replacing a composite solid propellant with a gel fuel will make the dynamic engine thrust control in time relatively easy to implement. In addition, gel fuels, compared to liquid fuels, have lower fire hazard indicators due to the minimized evaporation losses and leakage during storage. Components of gel fuels and products of their combustion are characterized, as a rule, by less negative impact on the environment compared to typical liquid rocket fuels [6].

Among the wide variety of condensed substances used as fuels, it is possible to distinguish conditionally: solid, liquid, gaseous and gel fuels. The latter are less common in

* Corresponding author: nigayalexandr@mail.ru

practice. The reason for this is the lack of a general theory of ignition and combustion of such fuels in comparison with solid, liquid and gaseous condensed substances [7]. Physical and chemical processes occurring during ignition and combustion of solid, liquid and gaseous fuels are well studied. Mathematical models [8] have been developed to reliably predict the main characteristics and to use these results in the design of engines and power plants to optimize their functioning. Developing the theoretical foundations of energy technologies for combustion of gel fuels is impossible without basic research of ignition and combustion processes. The need for such studies is explained by the advantages of gel fuels in comparison with other types of condensed substances. Due to their semi-solid and semi-liquid state, these fuels combine advantages of both solid and liquid propellants.

The aim of this research is to define the thermophysical characteristics of gel fuel under various heating conditions.

2 Experimental investigation

Fig. 1 shows the scheme of the experimental setup, the main part of which is a tubular muffle furnace Nabertherm. The temperature range is 300–1500 K. In each series of experiments, the ceramic tube was heated to a given temperature. After its stabilization, the metal particle was placed in the center of the ceramic tube and heated for 10 min to the ambient temperature of the high-temperature environment. Uniform temperature distribution in the particle was achieved [9].

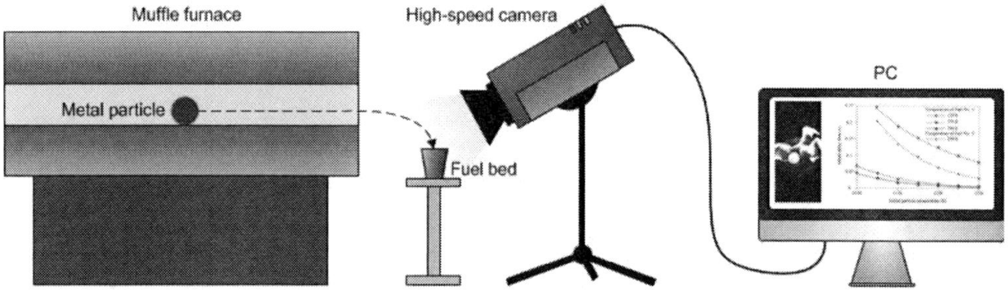

Fig. 1. The scheme of the experimental setup.

After that it was released to fall perpendicularly to the fuel surface into the metal vessel. The time of free fall of the hot particle in the air until the moment of contact with the fuel sample surface was about 50 ms.

The processes occurring during the induction period of the interaction of the hot particle with the fuel bed were recorded with high-speed camera Phantom v411. It characteristics are: 4200 frames per second at maximum resolution 1280×800 pixels; pixel size 20 μm; 12 bit depth. The exact contact time of the hot particle with the surface of fuel bed was automatically recorded by a high-speed camera. Trigger detected two moments: the contact between particle and fuel surface, and the flame appearance. The ignition delay time is the parameter of interest and is defined by the elapsed time between these two signals. The random error of delay time determination was calculated according to the results of the set of 10-15 experiments.

Experimental studies of the ignition process were performed at different fuel temperatures: 120, 260, and 290 K. Cooling of the fuel samples to minimum temperatures took place in several stages: preparation, freezer, liquid nitrogen and experiment.

The main characteristics of the fuel: density at 293 K 920 kg/m^3; humidity 0.28 % mass.; ash content 0.78 % mass.; ignition temperature 491 K.

3 Results

As a result of the performed experimental research the basic regularities and characteristics of gel fuel ignition at local heating have been established. This paper studied the physical and chemical processes occurring during the time period from the moment of a hot metal particle contact with the fuel surface to the moment of flame appearance in the vicinity of a local heating source. A metal particle with a characteristic size of 10 mm was deposited inertly on the fuel surface. The temperature of the samples of gel fuel in different experiments was 120, 260, and 290 K. At the initial fuel temperature of 290 K, the hot particle was embedded in its surface layer to a depth of 4–6 mm. At lower fuel temperatures of 260 K the metal particle was embedded into the fuel surface layer to a depth from 3 to 5 mm. At fuel temperatures close to cryogenic values, the hot particle was not embedded into the fuel surface layer (Fig. 2) as a result of inertial deposition. Under such conditions, the fuel is in the solid state.

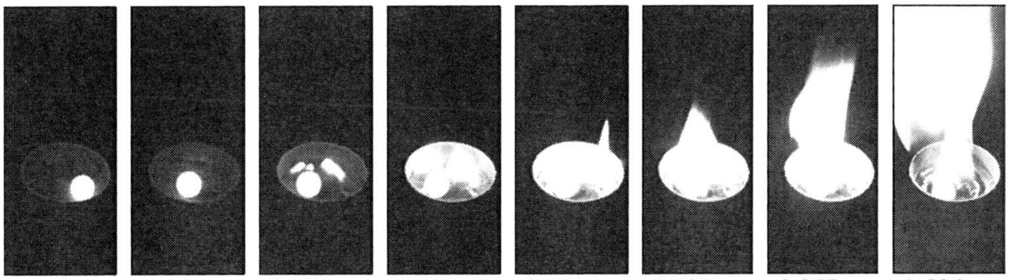

t=0 0.056 s 0.114 s 0.121 s 0.161 s 0.207 s 0.277 s 1.586 s

Fig. 2. Videogram of gel fuel ignition at an initial temperature of 120 K by a steel particle in the sphere shape of 10 mm in size at an initial temperature of 1350 K.

The ignition delay time of the gel fuel was set (Figs. 3, 4) at the variation of initial temperature of heating sources in the range of 1050–1350 K. Research was executed at different initial temperatures of fuel samples: 120, 260, and 290 K. Ignition sources were metal particles in the shape of sphere, disk, and parallelepiped with characteristic size of 10 mm.

It is established that under the conditions of the experiments, the initial temperature of the steel particle of 1050 K is the minimum temperature required for ignition of the gel fuel. At $T_p<1050$ K the fuel did not ignite. The energy of the local source was not enough for heating, melting, evaporation of the fuel, formation of a combustible gas mixture and its ignition. In this case, as a result of heating, the fuel melted and evaporated without further ignition. The particle was completely immersed in the liquid. The local heating source did not contact the combustible gas mixture. It did not heat up and, accordingly, did not ignite.

The found dependences $t_d=f(T_p)$ (Figs. 3, 4) are close to linear at variation of the initial temperature of the hot particle in a wide range of $T_p=1050$–1350 K. The intensity of the transition process increases by less than 30% with a greater heat content of the energy source, characterized by its shape and initial temperature. The results of the study allow concluding that the main characteristic of gel fuel ignition at local heating is the minimum temperature of the particle necessary for combustion initiation. The increase in the initial temperature of the local energy source when exceeding the threshold conditions for fuel ignition had a negligible effect on the reduction of the ignition delay time (Figs. 3, 4) at moderately low (200–290 K) initial fuel temperatures.

At initial fuel temperatures close to cryogenic, the duration of the induction period increases 3–5 times (Fig. 4) in comparison with the values of moderately low (200–290 K) temperatures. The obtained result is explained by rather significant influence of the stages of inert heating and fuel melting on the ignition delay times. In such conditions, large amounts of energy are required to initiate fuel combustion. Over time, due to heat transfer to the ambient medium, the temperature of the metal particle is significantly reduced. The cumulative effect of these interrelated processes is manifested in the increase of the ignition delay time.

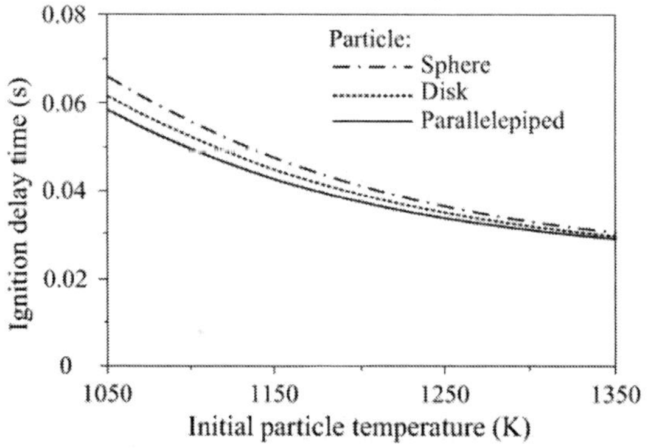

Fig. 3. Ignition delay times for a gel fuel (at T0=250 K) depending on the initial temperature of the steel particle in the shape of a sphere, a disk, or a parallelepiped of 10 mm in size.

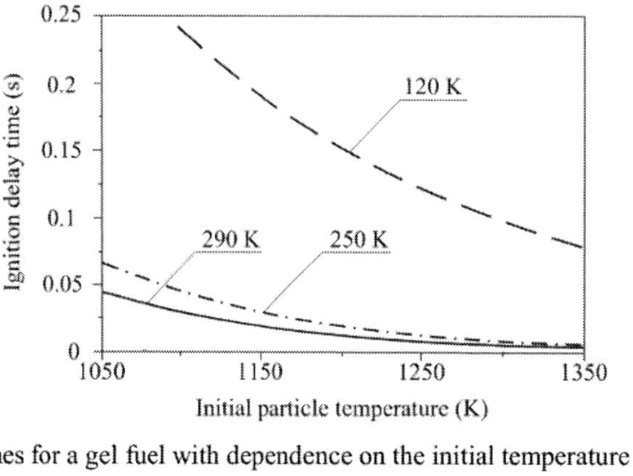

Fig. 4. Ignition delay times for a gel fuel with dependence on the initial temperature of the 10 mm spherical steel particle.

3 Conclusions

1. Small size metal particles heated to high temperatures can be used as a source of stable ignition of gel fuels, even at the cryogenic initial temperatures of the fuel (120 K).

2. The size and shape of the local heating source almost did not affect the characteristics of the fuel ignition in the case of stable ignition. The main factor is the heat content of the local energy source.

3. To realize stable ignition of gel fuel at 120 K it is necessary for the initial temperature of a local heating source to be larger than 1050 K at its characteristic size about 10 mm.

This work was supported by the Russian Science Foundation [grant number 18-13-00031].

References

1. M. Klick, Polar Record J. **52** (2016)

2. C. Lehnert, E. Karlsson, C. Giannopapa, Acta Astronautica J. **140** (2017)

3. Y. B. Zeldovich, G. I. Barenblatt, V. B. Librovich, G. M Makhviladze, *Mathematical Theory of Combustion and Explosions* (Plenum, New York, United States, 1985)

4. F. A. Williams, *Combustion Theory* (Westview Press, Boulder, United States, 1985)

5. V. I. Tsutsuran, N. V. Petrukhin, S. A. Gusev, *Military Engineering Analysis of the State and Prospects of Development of Propellants* (Ministry of Defense RF, Moscow, Russia, 1999) [in Russian]

6. T. M. Klapötke, *Chemistry of High-Energy Materials* (Walter de Gruyter GmbH, Berlin/Boston, Germany, 2015)

7. W. A. Sirignano, Progress in Energy and Combustion Science J. **42,** 1 (2014)

8. A. Krisman, E.R. Hawkes, M. Talei, A. Bhagatwala, J.H. Chen, Combustion and Flame J. **172** (2016)

9. A. Nigay, N. Gal'chenko, V. Medvedev, D. Sivkov, MATEC Web Conf. **72**, 4 (2016)

The Acoustic Spectrums of the Combustion Process in the IC-Engines

Ramil Shaymukhametov[1,*] and *Vladimir* Strekalov[1]

[1]Kazan Federal University, Kremlyovskaya St., 18, Kazan, 420008, Russia

Abstract. The results of the Fourier spectroscopy of the acoustic noise of an internal combustion engine are presented. The sound was recorded using a voice recorders placed near the four-cylinder petrol engine. The engine speed was set to the fixed values from 900 to 3000 rpm. The distinctive characteristics of the periodograms where associated with the certain physical processes in the engine.

1 Introduction

The flame in a combustion chamber is an effective generator of acoustic noise. The sound energy from combustion processes comprises more than 70% of the total noise energy in a modern car.

The generation of sound itself is a complex process. The level of a harmonic of the spectrum depends not only on an oscillator, but also on various resonators. Therefore, from the spectrum of sound, one can not judge the quantitative distribution of the dynamic elements of the flame. Nevertheless, a comparative analysis of the acoustic noise spectrums under various engine operating conditions has much potential for yielding information about the combustion.

2 Experiment technology and data processing

The measurements were carried out on engines of various cars (Hyundai i40, 2.0 L, 150 hp; Hyundai Accent, 1.6 L., 105 hp; Chevrolet Cruze 1.6 L, 109 hp; Peugeot 4007, 2.4 L, 170 hp; Opel Astra 1.6 L, 115 hp). The acoustic signals were recorded using a voice recorders installed under the car's hood near the engine cylinders (in the frequency range 200-7000 Hz, signal-to-noise ratio 37 dB). The sound reference generator being located near recorders was used for calibrating. The cars stood motionless in neutral gear. The engine speed was set to the fixed values from 900 to 3000 rpm. The raw data were collected at a sampling frequency of 16 kHz with 16-bits per sample. The spectral analysis of the data was carried out according to a standard procedure [2]. The frame length was 2048, each frame was 25% overlapped. Frames were windowed with Hamming window and thereafter FFT was applied to each frame.

* Corresponding author : ramil.shai@mail.ru

3 Experimental data and speculations

The energy spectrums have complex line structure. Low-frequency components are missing since the microphone bandwidth began at 200 Hz. The main energy of the acoustic noise is concentrated in the band 500 - 2500 Hz (Fig. 1).

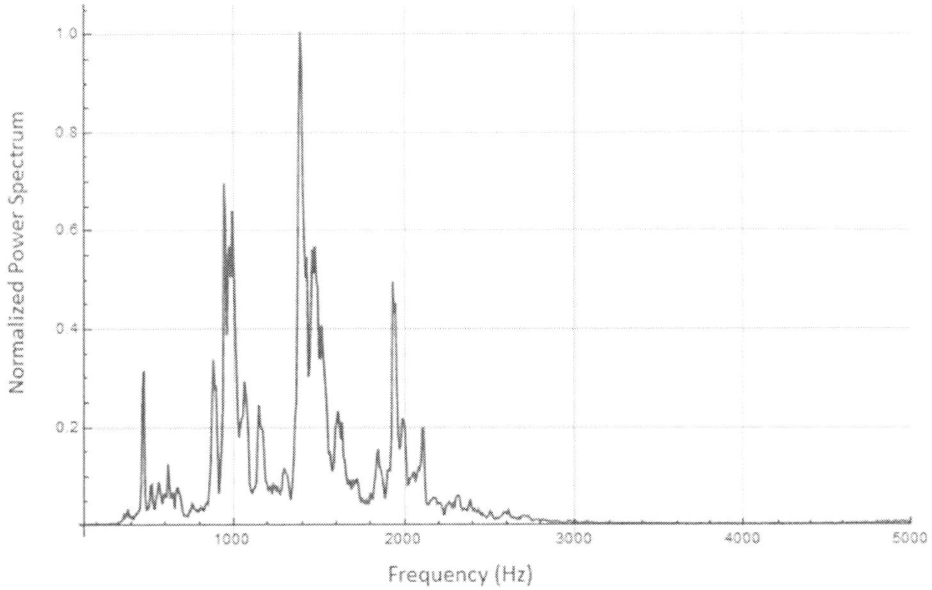

Fig. 1. Typical energy spectrum of engine noise. The crankshaft frequency is 20 Hz (1200 RPM).

The frequency of the maximum and spectrum width depend only weakly on the engine speed (Fig. 2). The main characteristics of the spectrums are not significantly different for various brands of cars with the same engine volume.

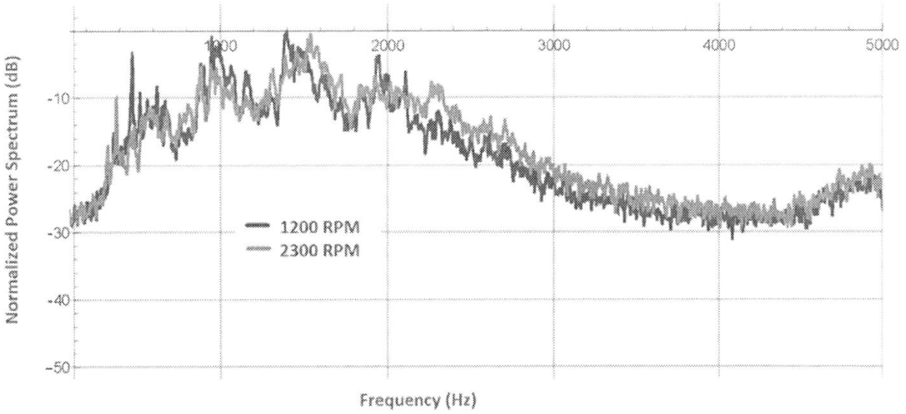

Fig. 2. Energy spectrums with a change in the frequency of rotation of the crankshaft.

Under the hood of the car a sound reference generator was installed to calibrate the recorders. The measurements of absolute values showed a strong dependence of the noise energy on the crankshaft rotation speed. The results are depicted in Fig. 3. When the engine speed is increased by a factor of 2, the sound noise energy is increased approximately 10 times.

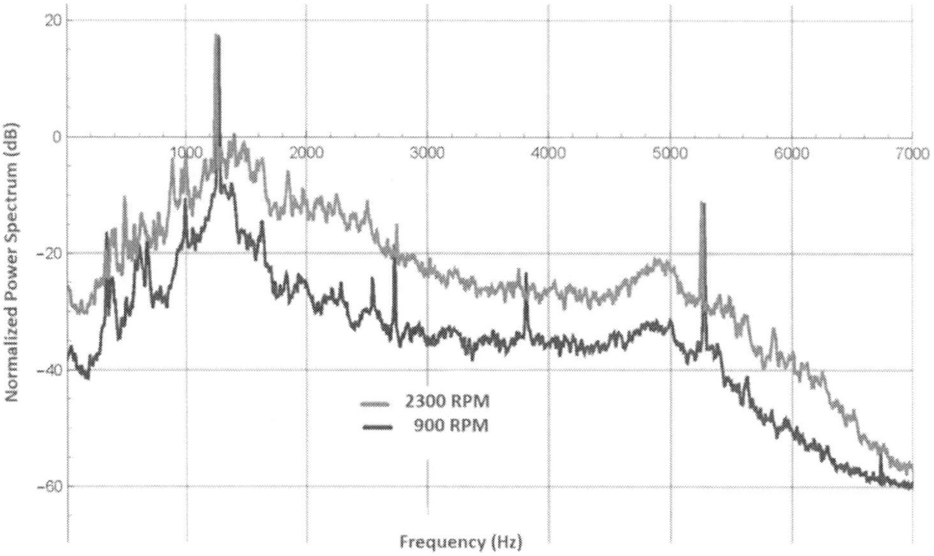

Fig. 3. The dependence of the power level on the frequency of rotation of the crankshaft.

It is likely that a cylinder volume governs the width and structure of the spectrum. Based on the flame propagation velocity in piston engines in the absence of detonation of 50-80 m/s , and the characteristic size of the combustion chamber 0.1 m (cylinder diameter), a lower frequency of combustion noise of 500-800 Hz can be estimated. These evaluations confirmed by observations at different volumes motor (Fig. 4). As the volume of the engine cylinder increases, the spectrum of sound becomes narrower and the distance between the peaks decreases.

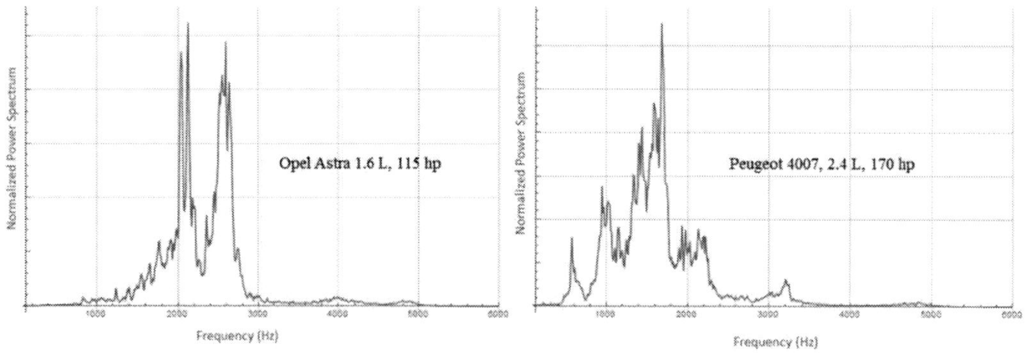

Fig. 4. Spectrums of acoustic noise at different engine volumes.

3 Concluding remarks

The generation of sound is determined by complex nonlinear mechanisms, and therefore it is impossible to make a quantitative description of the combustion processes in the engine from the acoustic noise spectrum. However, comparing the spectrums under different observation conditions, one can draw qualitative conclusions about combustion as a source of sound.

The weak dependence of the spectrum characteristics on the engine speed shows that it is the combustion process that is the main source of the sound. The movements of the mechanical components of the motor only modulate the sound signal.

In the high-frequency part of the spectrum (above 4000 Hz), an increased level of energy is noticeable (Fig. 4), which can be explained by the presence of detonation combustion waves. The velocity of the flame front during a detonation exceeds the speed of sound, so the detonation can generate the noise in the high-frequency part of the spectrum.

Experimental records of engine acoustic noises contain a variety of information about the dynamics of the combustion. Determination of the dependence of the spectrum characteristics on the cylinder's geometric dimensions and fuel quality is of great practical importance for the development of internal combustion engines.

References

1. R. Hickling, M.M. Kamal, Engine Noise: Excitation, Vibration, and Radiation, first ed. Springer, (1982)
2. Bendat, J. S. and Piersol, A. G. Engineering applications of correlation and spectral analysis, John Wiley & Sons.M.(1980)
3. Withrow L, Rassweiler GM Slow motion shows knocking and nonknocking explosions. *SAE Transactions* **39**, 297-303, 312 (1936)
4. Heywood JB. Internal combustion engine fundamentals. New York: McGraw-Hill; (1988)
5. D.E. Baxa, Noise Control in Internal Combustion Engines, Krieger Pub Co, (1989)

The reaction of 2-naphthyl with 1,3-butadiene: a theoretical study

Oleinikov Artem[1], *Mebel* Alexander[1,2*], and *Azyazov* Valeriy[1,3]

[1]Samara National Research University, 443086, Samara, Russian Federation
[2]Florida International University, FL 33199, Miami, USA
[3]Lebedev Physical Institute, 443011, Samara, Russian Federation

Abstract. This work is devoted to a theoretical study of the 2-naphthyl + 1,3-butadiene reaction aimed at understanding the hydrogen-loss mechanism in the reactions of n-naphthyl (n=1,2) with 1,3-butadiene. All intermediates and transition states, their vibrational frequencies and zero-point vibrational energies, and the potential energy diagram were calculated at the G3(MP2,CC)//B3LYP/6-311G** theoretical level. The estimate of the reaction rate constant gives the order of 10^{-13} cm^3 $molecule^{-1}$ s^{-1} at temperatures below 500 K.

1 Introduction

A great amount of theoretical and experimental research in the field of polycyclic aromatic hydrocarbons (PAH) growth and soot formation have been reported in last decades. These investigations demonstrated harmful effects of PAH on the environment and human health [1-3]. Due to high thermal stability of aromatic radicals and resonantly stabilized free radicals the following mechanisms are considered important in the PAH and soot precursor formation: hydrogen abstraction–acetylene addition (HACA) [4–6], phenyl addition–cyclization [7, 8], and ethynyl addition [9]. However, more complex PAH such as three-ring phenanthrene and anthracene could not be formed through HACA-based mechanisms. In this article, the addition of a 1,3-butadiene molecule is proposed as a key step in the formation of a third aromatic ring.

Recently obtained experimental and theoretical results on the reaction of 1-naphthyl with 1,3-butadiene defy the thesis that PAH growth is predominantly a high temperature phenomenon [10] and a further support for this conclusion is provided here for the analogous reaction of 2-naphthyl.

2 Methods

Optimization of geometry of reactants, products, intermediates, and transition states, and zero-point vibrational energy calculations were performed by using DFT calculations at the B3LYP/6-311G** level. Single point energies were refined by using higher level of theory methods and calculated using the following scheme: E_0 [G3(MP2, CC)] = E

* Corresponding author: mebela@fiu.edu

[CCSD(T) /6 − 311G**] + ΔEMP2 + E(ZPE), where ΔEMP2 = E[MP2/G3Large] − E[MP2/6 − 311G**] is a basis set correction and E(ZPE) is the zero-point energy. The calculations were performed by using the GAUSSIAN 09 [2] and MOLPRO 2010 packages [3].

Reaction rate constants for all reaction steps were computed by solving one-dimensional master equation using Rice-Ramsperger-Kassel-Marcus theory. The harmonic approximation was applied to calculate partition functions using B3LYP/6-311G** frequencies.

3 Results

Quantum chemical calculations at the theory level described above provide information about five hydrogen loss channels in the bimolecular reaction of 2-naphthyl with 1,3-butadiene. All the investigated species and intermediates are represented in Figure 1 with relative energies given in kcal mol^{-1} with respect to the reactants. The first channel leading to P4 + H products starts with a small barrier (TS0-4) 3.87 kcal mol^{-1} above the reactants. This is a formation of a bond between the radical site in 2-naphthyl and a second C atom in 1,3-butadiene leading to intermediate W4 located in a potential well 32.78 kcal mol^{-1} below the reagents. The further development of this channel passes through a hydrogen loss barrier TS4-P4 from the attacked carbon in 1,3-butadiene. The barrier height is 32.59 kcal mol^{-1} and the channel forms the P4 species that is 7.36 kcal mol^{-1} lower in energy than the separated reagents. The initial reactants also have a second channel leading to the P1, P2, P3 and P6 products presented in Figure 1. This channel starts with a van der Waals complex W0 located at -1.72 kcal mol^{-1} that represents a barrier-free entry into the reaction channel. After that a bond is formed between the radical site in 2-naphthyl and the terminal carbon atom of 1,3-butadiene via the transition state TS0-1 (-1.07 kcal mol^{-1}) descending into a potential well W1 at -46.82 kcal mol^{-1}. Hydrogen loss from C1 in 1,3-butadiene represents the reaction channel from W1 to the P1 + H products (at -8.98 kcal mol^{-1}) via the transition state TS1-P1 located at -5.91 kcal mol^{-1}. The reaction can also continue from W1 (-46.82 kcal mol^{-1}) to W2 (-46.26 kcal mol^{-1}) intermediates that are close in energy and the linking transition state TS1-2 at -32.67 kcal mol^{-1} is a trans-cis rearrangement in the 1,3-butadiene side chain. The next endoergic hydrogen loss from C1-carbon in butadiene pathway leads to the P2 + H products at -6.04 kcal mol^{-1} via transition state TS2-P2 at -2.95 kcal mol^{-1}. Other two pathways lead to P3 (1,4-dihydrophenanthrene) and P6 (1,4-dihydroanthracene). Both mechanisms involve ring closure. The channel to P3 passes via TS2-3 (-27.29 kcal mol^{-1}) to W3 (-52.23 kcal mol^{-1}) with ring closure on the C1-carbon of the former 2-naphthyl reactant and after that proceeds via hydrogen loss from the C1-carbon via transition state TS3-P3 (-25.26 kcal mol^{-1}) and gives the product P3 + H at -25.05 kcal mol^{-1}. The last reaction channel has the similar mechanism but the ring closure in TS2-6 (-26.24 kcal mol^{-1}) occurs on the C3-carbon in 2-naphthyl. After that reaction goes through intermediate W6 (-46.97 kcal mol^{-1}) and hydrogen-loss transition state TS6-P6 (-18.11 kcal mol^{-1}) and finds its end in the product P6 + H located at -24.26 kcal mol^{-1}.

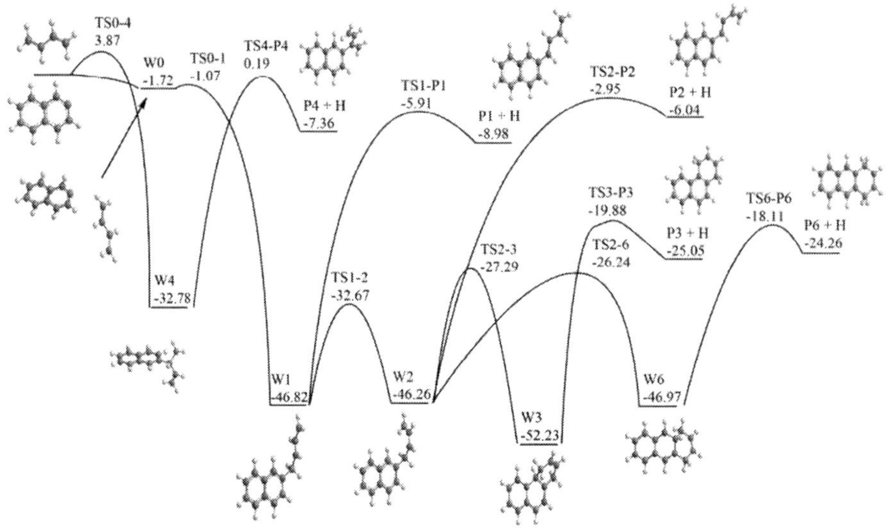

Fig. 1. Potential energy diagram for the reaction of 2-naphthyl with 1,3-butadiene; the energies are given in kcal mol^{-1} relative to the separated reactants.

RRKM-Master Equation calculations were exploited to obtain reaction rate constants for five pressures indicated in Fig. 2 in the range of temperature from 300 to 2500 K. The calculated temperature-dependent reaction rate constants for pressures 0.01, 0.1, 1, 10, 100 atm and the high pressure limit are presented in Figure 2. The curves behaviour below 700 K is similar and coincides with the HP limit. Some small decrease could be seen for all finite pressures at higher temperatures and after that the rate constants merge in one curve again.

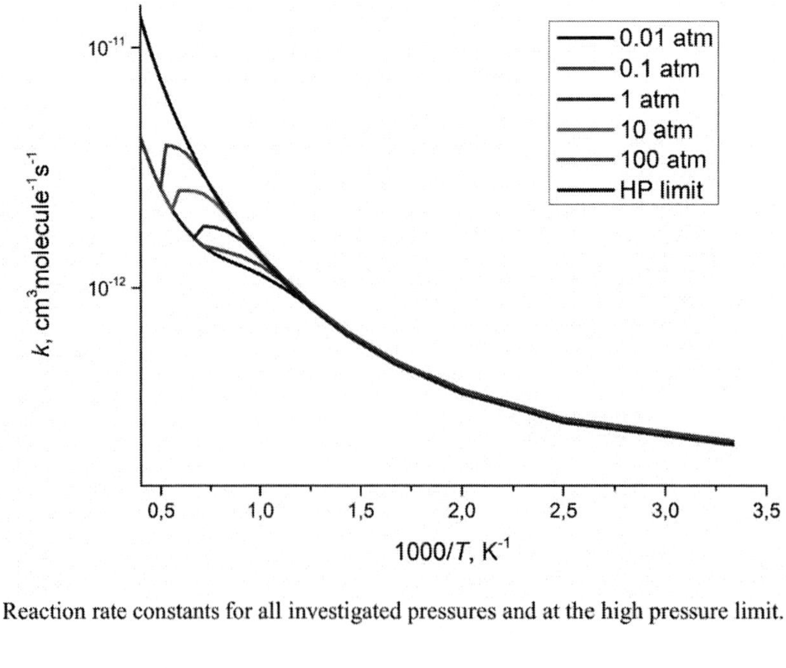

Fig. 2. Reaction rate constants for all investigated pressures and at the high pressure limit.

The analysis of the branching ratios allows us to conclude that the channels leading P1 and P2 plus H would be the main channels in the reaction at high temperatures, whereas the three-ring products dihydrophananthrene P3 and dihydroanthracene P6 show significant yields in the 1200-2000 K range. The reaction shows low pressure dependence and the product branching ratios for P1-P6 are presented in Table 1. For the temperatures below 1200 K, W1 and W2 complexes get collisionally stabilized and the resulting contributions are presented in Table 2.

Table 1. Branching ratios of the reaction of 2-naphthyl with 1,3-butadiene for pressure 1 atm.

T, K	P1	P2	P3	P4	P6
1200	3,08%	1,44%	4,13%	0,02%	2,93%
1300	28,89%	14,39%	32,61%	0,27%	22,87%
1400	35,98%	19,47%	25,38%	0,62%	18,56%
1500	41,30%	24,01%	19,28%	0,81%	14,60%
1600	45,09%	27,91%	14,61%	1,00%	11,38%
1700	47,64%	31,16%	11,14%	1,17%	8,89%
1800	49,26%	33,82%	8,60%	1,33%	6,99%
1900	50,25%	36,00%	6,72%	1,47%	5,56%
2000	50,79%	37,79%	5,34%	1,60%	4,48%
2100	51,05%	39,27%	4,30%	1,72%	3,65%
2200	51,13%	40,51%	3,51%	1,84%	3,01%
2300	51,10%	41,54%	2,91%	1,94%	2,51%
2400	51,01%	42,40%	2,44%	2,04%	2,12%
2500	50,88%	43,12%	2,06%	2,13%	1,80%

Table 2. Branching ratios of the W1 and W2 intermediates at the low temperatures for 1 atm.

T, K	W1	W2
300	34,91%	65,05%
400	33,36%	66,49%
500	32,73%	66,79%
600	29,82%	69,08%
700		98,21%
800		97,72%
900		97,40%
1000		96,30%
1100		93,76%
1200		88,25%

4 Conclusion

The mechanism of the 2-naphthyl + 1,3-butadiene reaction has been proposed in this study. The potential energy diagram and reaction rate constants with branching ratios were obtained in a broad set of pressures and temperatures. This study is important to complement the assumption PAH generation is not only a high-temperature phenomenon. P1 and P2 with hydrogen are concluded to be the main products at high temperatures but the three-ring products dihydrophenanthrene and dihidroanthracene can also be readily formed at certain conditions. The calculated reaction rate constants show low pressure dependence at pressures above 0.01 atm.

The work was supported by the Ministry of Education and Science of the Russian Federation under the Grant No. 14.Y26.31.0020 to Samara University.

References

1. B.S. Haynes, H.G. Wagner, Prog. Energy Combust. Sci. **7** (1981).
2. M. Frenklach, Phys. Chem. Chem. Phys. **4** (2002).
3. H. Richter, J.B. Howard, Prog. Energy Combust. Sci. **26** (2000).
4. M. Frenklach, E. D. Feigelson, Astrophys. J. **341** (1989).
5. M. Frenklach, H. Wang, Symp. (Int.) Combust. **23** (1991).
6. I. Cherchneff, J. R. Barker, A. G. G. M. Tielens, Astrophys. J. **401** (1992).
7. B. Shukla, M. Koshi, Combust. Flame **158** (2011).
8. B. Shukla, K. Tsuchiya, M. Koshi M. 2011. J. Phys.Chem. A **115** (2011).
9. A. M. Mebel, V. V. Kislov, R. I. Kaiser. 2008 J. Am. Chem. Soc. **130** (2008).
10. A. M. Thomas, M. Lucas, T. Yang, R. I. Kaiser, L. Fuentes, D. Belisario-Lara, A. M. Mebel, ChemPhysChem **18**, 15 (2017).
11. M.J. Frisch et al. *GAUSSIAN 09*, Revision A.1 (Gaussian, Inc., Wallingford, CT, 2009).
12. *MOLPRO,* version 2010.1, a package of ab initio programs, H.J.Werner and others, see http://www.molpro.net.

Study of a dielectric barrier discharge burner for plasma assisted combustion

Alena Sludnova[1,2], *Nikolay* Mironov[1], and *Paul* Mikheyev[1,2*]

[1]Samara National Research University, 443086, Samara, Russia
[2] P N Lebedev Physical Institute, 443011, Samara, Russia

Abstract. In this short communication we report the results of measurements of the number densities of O_3, NO and NO_2 in the afterglow of a dielectric barrier discharge. Detailed modelling of plasma processes in a DBD is a challenging task and this data is necessary for verification of numerical model. The DBD in air produced only [O_3] up to ~10^{16} cm^{-3} and [NO] up to ~5×10^{14} cm^{-3} that increased with discharge power linearly. In air fuel mixtures [O_3] was an order of magnitude smaller, it decreased with [CH_4]. In methane-air mixture NO was not detected, but [NO_2] up to ~10^{15} cm^{-3} appeared instead.

1 Introduction

Low temperature nonequilibrium plasma is an efficient tool for speeding up chemical processes relative to combustion, and extensive studies of plasma ignition and plasma-assisted combustion are under way at present [1, 2]. For example, a repetitively pulsed nanosecond discharge with a low duty cycle [3] in an air-fuel flow permitted to achieve ignition, flame holding and steady combustion downstream of the discharge at low pressure of 60 Torr. The reduced electric field *E/N* in this type of discharge is on the order of hundreds of Td and large numbers of energetic electrons efficiently dissociate, ionize and excite gas molecules. However, with pressure one needs to increase a discharge current substantially to fill the discharge gap with plasma. Therefore, at larger pressures, scaling of this type of discharge does pose technical problems due to large pulsed discharge current.

Dielectric barrier discharge (DBD) is of special interest in this field, because it is able to provide plasma with a large *E/N*, too, even at an atmospheric pressure. It operates in a wide pressure range, quite simple to design, has an ability to be easily integrated in different flow configurations. At an atmospheric pressure, its current density is on the order of 100 mA cm^{-2} and voltage across the electrodes with 1 mm gap in air-fuel mixture is about 6 kV. DBD consists of a large number of streamers of a nanosecond duration with a large *E/N* in their heads and is able to ionize, dissociate and excite molecules efficiently. Detailed modeling of plasma processes in a DBD that lead to ignition of an air-fuel mixture is a challenging task, but it can be simplified if we try to calibrate the model first, measuring the number density of products that appear in the plasma afterglow before ignition.

* Corresponding author: paulmikheyev@hotmail.com

The aim of this work was to measure O_3, NO and NO_2 number densities in the afterglow of a DBD in an air-methane mixtures with intention to use this data for verification of numerical modeling.

2 Experiment

The sketch of the experimental setup is shown in Figure 1. A diffuse DBD was obtained in a coaxial burner. The burner was made out of a quartz tube 16 mm inner diameter and 1.5 mm wall thickness. The gap between the inner anodized aluminum electrode and the tube wall was 1 mm. An outer electrode was a fine steel mesh 5 mm long along the tube wall, providing a homogeneous discharge in air [4] and CH_4:Air mixtures.

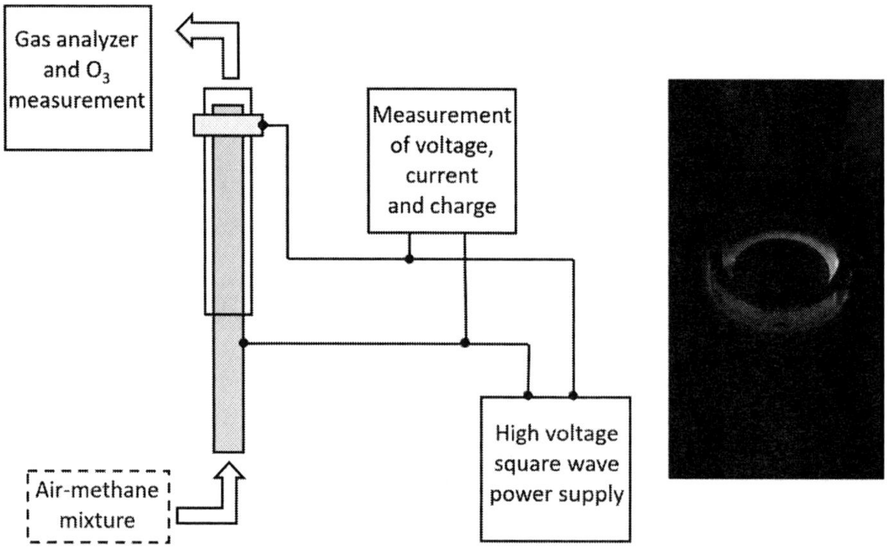

Fig. 1. Experimental setup and a photograph of the DBD from above at an angle.

Voltage, applied to the electrodes, discharge current and transferred charge were measured with the help of a voltage divider, a Rogowski coil and a capacitor in series with the discharge. Discharge power was determined by calculating the area of Lissajous figures in the voltage-charge coordinates, which is a standard technique for a DBD. Discharge started at ~6.5 kV. Power load into the discharge was on the order of 10 W. Small discharge gap prevented ignition of CH_4:Air mixtures.

Ozone number density was measured by absorption of 255 nm radiation from a LED. Commercially available gas analyzer GMS-810 measured the number densities of nitrogen oxides downstream of a cell for O_3 measurement.

3 Results and discussion

In dry air the ozone number density [O_3] was on the order of 10^{16} cm^{-3}, as represented in figure 2, and exhibited a linear dependence on discharge power.

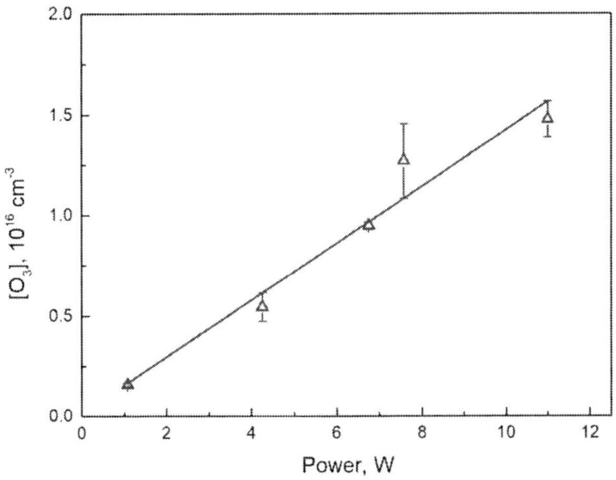

Fig. 2. Ozone number densities in dry air.

In methane-air mixtures [O_3] was only on the order of 10^{15} cm^{-3}, decreased with methane content and exhibited linear dependence on discharge power, too, as shown in figure 3.

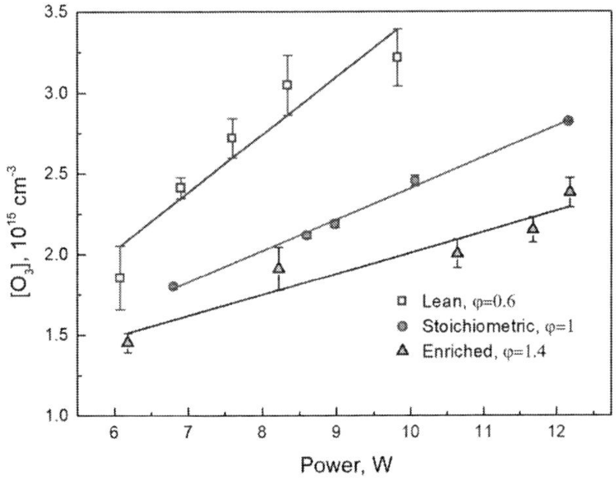

Figure 3. Ozone number densities for different methane-air mixtures.

Measurements of the number densities of nitrogen oxides for air revealed that DBD in air produced only O_3 and NO that increased with discharge power linearly. Apparently, NO_2 was beyond the detection limit of the gas analyzer.

In stoichiometric methane-air mixture [O_3] was an order of magnitude smaller, NO was not detected, but NO_2 appeared instead. The results of NOx measurements are in figure 4.

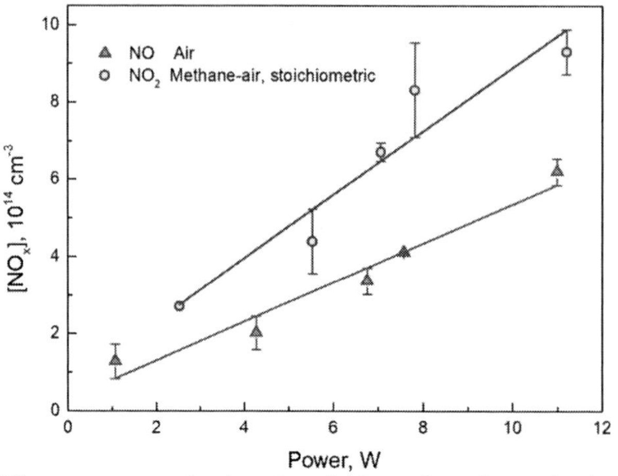

Figure 4. Results of NOx measurements in air an in stoichiometric methane-air mixture.

Qualitatively, our NOx measurements agree with the results reported in [5, 6].

This work was supported by the Ministry of Education and Science of the Russian Federation under the Grant No. 14.Y26.31.0020 to Samara University.

References

1. A. Starikovskiy, N. Aleksandrov, *Progress in Energy and Comb. Sci.* **39**, 61 (2013)
2. S. M. Starikovskaia, *J. Phys. D: Appl. Phys.* **47**, 353001 (2014)
3. A. Bao, Y. G. Utkin, S. Keshav, G. Lou, I. V. Adamovich, *IEEE Trans. on Plasma Sci.* **35**, 1628 (2007)
4. S. Okazaki, M. Kogoma, M. Uehara and Y. Kimura, *J. Phys. D: Appl. Phys.* **26**, 889 (1993)
5. T. Namihira, S. Tsukamoto, D. Wang, S. Katsuki, H. Akiyama, R. Hackam and K. Okamoto, "Pulsed Power Conference, 1999. Digest of Technical Papers. 12th IEEE International," in *Production of nitric monoxide in dry air using pulsed discharge*, Monterey, (1999)
6. N. Rehbein and V. Cooray, *Journal of Electrostatics* **51** 333, (2001)

AUTHOR INDEX